Quality Control of Gamma Cameras and Nuclear Medicine Computer Systems

IPEM Report Number 111

A Report produced by the Institute of Physics and Engineering in Medicine

Editor
Ewan Eadie, Dundee

Authors
Ian Armstrong, Manchester
Mike Avison, Bradford
Natalie Bebbington, Birmingham
Philip Cosgriff, Boston
Ewan Eadie, Dundee
Glen Gardner, Dundee
Brian Hutton, London
Joanne Kerry, Lincoln
Richard Lawson, Manchester
David J. Platten, Northampton
Therese Soderlund, Singapore
Penelope J. Thorley, Leeds
Fred Wickham, London

Contributors

Ian Armstrong
Nuclear Medicine, Central Manchester University Hospitals

Mike Avison
Medical Physics, Bradford Royal Infirmary, BD9 6RJ

Natalie Bebbington
Nuclear Medicine, Queen Elizabeth Hospital Birmingham

Philip Cosgriff
Nuclear Medicine Department, Pilgrim Hospital, Boston

Ewan Eadie
Nuclear Medicine, Ninewells Hospital and Medical School, Dundee

Glen Gardner
Nuclear Medicine, Ninewells Hospital and Medical School, Dundee

Brian Hutton
Institute of Nuclear Medicine, University College London and UCLH Foundation Trust, London

Joanne Kerry
Nuclear Medicine, Lincoln County Hospital, ULHT NHS Trust

Richard Lawson
Central Manchester Nuclear Medicine Centre, Manchester

David J Platten
Medical Physics Department, Northampton General Hospital

Therese Soderlund
A*STAR-NUS Clinical Imaging Research Centre, Singapore

Penelope J Thorley
Medical Physics, Leeds Teaching Hospitals, Leeds

Fred Wickham
Nuclear Medicine Department, Royal Free London NHS Foundation Trust

Fairmount House, 230 Tadcaster Road
York YO24 1ES
ISBN 978-1-903613-60-3

Published by the Institute of Physics and Engineering in Medicine
Fairmount House, 230 Tadcaster Road, York YO24 1ES

Prepared by:
The Charlesworth Group, Huddersfield, UK. www.charlesworth-group.com

Contents

List of abbreviations used

1D	one-dimensional
2D	two-dimensional
3D	three-dimensional
^{201}Tl	Thalium-201
^{57}Co	Cobalt-57
^{67}Ga	Gallium-67
^{99m}Tc	Technetium-99m
AAPM	American Association of Physicists in Medicine
ARSAC	Administration of Radioactive Substances Advisory Committee
AXrEM	Association of Healthcare Technology Providers for Imaging, Radiotherapy and Care
BIR	British Institute of Radiology
BMI	Body Mass Index
BNMS	British Nuclear Medicine Society
BSI	British Standards Institution
CE	Conformité Européenne
CFOV	central field of view
COR	centre of rotation
CoV	coefficient of variation
cps	counts per second
CsI	Caesium Iodide
CT	Computed Tomography
CTDI	Computed Tomography Dose Index
CUSUM	Cumulative Sum
CZT	Cadmium Zinc Telluride
DICOM	Digital Imaging and Communications in Medicine
DLP	dose length product

DMSA	dimercaptosuccinic acid
DRL	diagnostic reference level
EANM	European Association of Nuclear Medicine
EC	European Commission
ECG	electrocardiogram
EoS	end of service
EU	European Union
FBP	filtered back-projection
FOV	field of view
FWHM	full width at half maximum
FWTM	full width at tenth maximum
GBq	gigabecquerel
GE	General Electric
GFR	glomerular filtration rate
HPA	Health Protection Agency
HU	Hounsfield Units
IAEA	International Atomic Energy Agency
ICSIG	Informatics and Computing Special Interest Group
IEC	International Electrotechnical Commission
IPEM	Institute of Physics and Engineering in Medicine
IPR	intellectual property rights
IPSM	Institute of Physical Sciences in Medicine
IR(ME)R	Ionising Radiation (Medical Exposure) Regulations
IRR	Ionising Radiation Regulations
ISCORN	International Scientific Committee on Radionuclide Nephro-urology
ISO	International Organisation for Standardisation
IT	information technology
ITU	intensive therapy unit

Kcounts	kilocounts
kcps	kilocounts per second
keV	kiloelectron volt
kg	kilogram
kW	kiloWatt
LAN	local area network
LEGP	low-energy general purpose
LEHR	low-energy high-resolution
LSF	line spread function
LVEF	left ventricular ejection fraction
M	mean
MAG3	mercaptoacetyltriglycine
MARS	Medicines (Administration of Radioactive Substances) Regulations
MBq	megabecquerel
MDA	Medical Devices Agency
MES	managed equipment service
mGy	milliGray
MHRA	Medicines and Healthcare Products Regulatory Agency
MPE	Medical Physics Expert
MRI	magnetic resonance imaging
MSC STP	Modernising Scientific Careers Scientist Training Programme
MTF	modulation transfer function
MUGA	multi-gated acquisition
NaI	Sodium Iodide
NEMA	National Electrical Manufacturers Association
NHS	National Health Service
NHSLA	NHS Litigation Authority
NHSSC	NHS Supply Chain

NIH	National Institute of Health
NM	Nuclear Medicine
NMSIG	Nuclear Medicine Special Interest Group
NPL	National Physical Laboratory
NRD	National Reference Dose
NRPB	National Radiological Protection Board
OJEU	Official Journal of the European Union
OS-EM	ordered subsets expectation maximisation
PACS	Picture Archiving and Communications System
PDP	personal development plan
PET	Positron Emission Tomography
PFI	private finance initiative
PM	preventative maintenance
PMT	photomultiplier tube
PSF	point spread function
PTFE	polytetrafluoroethylene
QA	quality assurance
QC	quality control
QGS	quantitative gated SPECT
QPS	quantitative perfusion SPECT
R&D	research and development
ROI	region of interest
RPA	Radiation Protection Advisor
RPS	Radiation Protection Supervisor
SD	standard deviation
s	second(s)
SiPD	silicon photodiode
SNMMI	Society of Nuclear Medicine and Molecular Imaging
SPC	statistical process control

SPECT	Single Photon Emission Computed Tomography
TAC	time–activity curve
UFOV	useful field of view
UK	United Kingdom
UPS	uninterruptable power supply
US	ultrasound
USA	United States of America
VPN	virtual private network
WAN	wide area network

1 Background

Ewan Eadie

1.1 Introduction

This report has been prepared for the Nuclear Medicine Special Interest Group (NMSIG) of the Institute of Physics and Engineering in Medicine (IPEM). It replaces IPEM Report 86 published in 2003 (IPEM, 2003a). Since 2003, there have been major and rapid advances in technology especially with regard to the now routine provision of combined modality imaging in single photon emission computed tomography/computed tomography (SPECT/CT) and the wider availability of positron emission tomography/computed tomography (PET/CT). Many of the technological advances in the past decade have been in computing power and capabilities. As such, software and the way we interact with software, have evolved. Chapter 8 is no longer focused on in-house-developed software. An initial foray into general software quality assurance proved that the topic was too complex to be covered within a single chapter of this report. As the topic of in-house-developed software is cross-discipline, the Informatics and Computing Special Interest Group (ICSIG) of IPEM will address all of the issues of in-house-developed software in a separate future publication. In this report, Chapter 8 looks specifically at *Testing of Commercial Nuclear Medicine Software* providing practical advice on when and how to perform quality assurance checks.

In 2003, dedicated systems for PET were few and far between in the UK. Manufacturers were marketing cheaper coincidence options that could be added to a dual headed gamma camera without collimators, to make a crude gamma camera PET system. IPEM Report 86 (IPEM, 2003a) therefore included a chapter on *Quality Control for Gamma Camera PET Systems* but this has been removed from the current report as these systems are no longer considered to be relevant. Quality control (QC) of dedicated PET/CT systems is now covered in IPEM Report 108 *Quality Assurance of PET and PET/CT Systems* (IPEM, 2013). Similarly, a chapter in IPEM Report 86 on *Principles of Operation* has also been removed from this revision. The level of detail required to adequately describe gamma camera function is more comprehensively covered in a new IPEM book – The *Gamma Camera: A Comprehensive Guide* (Lawson, 2013).

Significant revision of the *Assessment of Planar Performance* (Chapter 3) has occurred as the chapter in the previous report had not been updated since the Institute of Physical Sciences in Medicine Report 66 published in 1992 (IPSM, 1992). In addition, *Whole-body Imaging Quality Control* (Chapter 4) and *SPECT Quality Control* (Chapter 5) have been revised with important additions.

There have been a further three important chapters added to this latest revision. Chapter 6 introduces the quality control checks required for gamma cameras with CT capability. It focuses on the tests that are unique to the dual modality systems and the CT-only tests that might be reasonably expected of a Nuclear Medicine department. More detailed testing of the CT component is described elsewhere, for example, in IPEM Report 91: Recommended Standards for the Routine Performance Testing of Diagnostic X-Ray Systems (IPEM, 2005).

Chapter 7 introduces novel gamma camera technology that, whilst relatively rare in the UK at the time of publication, may become more established as the evidence base increases and costs decrease. Several authors have contributed to this chapter to provide their experience of quality control of these new systems.

Finally, an important addition to this revision is Chapter 9, the *Life Cycle of a Gamma Camera*. Whilst each preceding chapter deals, in-depth, with the procedures and theory behind quality control tests, Chapter 9 places the tests within the context of a gamma camera's life-span. The QA process begins before a gamma camera purchase and so Chapter 9 begins with equipment specification and the tendering process. This chapter then follows the gamma camera as it progresses from installation and acceptance testing, to setting action thresholds for regular QC, and finally ending with recommendations on replacement and decommissioning of the gamma camera. In Chapter 9, the reader will also find a useful timetable of quality control testing, which summarises each test and its recommended frequency.

This report is a comprehensive guide to quality control of gamma camera systems and should be used in conjunction with the additional information referenced. It is intended to be a practical report that provides the reader with the means to carry out thorough quality control testing of their gamma camera.

2 Requirement for quality control

Philip Cosgriff and Richard Lawson

2.1 Introduction

A distinction can be made between the terms 'quality assurance' and 'quality control'. Quality assurance (QA) is an 'umbrella' term referring to all aspects of a procedure that contribute to the quality of the final product. In nuclear medicine, this would include staff training, testing of radiopharmaceuticals, assessment of equipment performance, and reporting of clinical studies. Quality control (QC) refers to the specification, assessment, optimisation, and maintenance of a particular aspect of the procedure, such as the performance of the imaging equipment.

2.2 Performance measurements

Gamma camera performance measurements may be divided into two categories. In the first category, the aim is to define performance, and to compare different systems. Standards of this type have been defined by the National Electrical Manufacturers Association (NEMA) in the USA (NEMA, 2012), the British Standards Institution (BSI) in the UK (BSI, 2005) and the International Electrotechnical Commission (IEC) in Europe (IEC, 2005), although the latter two are now identical through a harmonisation process.

Manufacturers usually quote the performance characteristics of their gamma cameras in terms of the NEMA specification, and this provides a convenient means of comparing gamma cameras when purchase is being considered. When a new camera has been installed, it is therefore necessary to confirm that the specified performance has been met using the NEMA methods as part of acceptance testing, and the European Association of Nuclear Medicine (EANM) has produced suitable guidelines for acceptance testing (Busemann Sokole et al., 2010a).

In the second category of performance measurements, the intention is to assess whether there has been any deterioration in the performance of the gamma camera since it was installed. While there is a significant overlap in the type of parameter which may be used for each kind of assessment, the main distinction is in the time and facilities required to carry out the measurements. Camera performance checks need to be done regularly, and some of them frequently, perhaps daily. They are carried out in a hospital environment, usually without access to the specialised test equipment available to manufacturers. In general, the NEMA methods do not lend themselves to such routine use, and indeed were not developed for this purpose.

The EANM has recommended a range of tests that should be carried out at regular intervals for routine quality control of gamma camera systems (Busemann Sokole et al., 2010b), though these guidelines do not include details of how the tests are to be performed. However, the International Atomic Energy Agency (IAEA) have published a more practical guide on how these tests may be carried out (IAEA, 2009) and there is a useful accompanying atlas illustrating a variety of camera fault conditions (IAEA, 2003).

This report builds on the work of previous IPEM reports on gamma camera quality control (HPA, 1978; IPSM, 1992; IPEM, 2003a). It describes practical tests that can be carried out routinely in the hospital environment to assess the performance of a gamma camera for routine clinical imaging. The tests should be within the capability of most nuclear medicine departments with support from the local medical physics department. The NEMA tests carried out as part of acceptance testing serve as baseline or reference tests, should the performance of the camera subsequently need to be traced back to its original performance. The QC tests described in this report should also be performed immediately following the acceptance tests. An important cross reference or 'bridge' is thus established between acceptance tests and routine QC tests at a time point when the system performance should be optimum. Thereafter, the QC tests should be repeated at the intervals indicated in this report (Section 9.6.1) to verify that the camera is still performing adequately for clinical use. Gamma camera quality control should thus be seen as part of a comprehensive preventative maintenance programme, in which unplanned down time is reduced to a minimum by careful monitoring of key aspects of system performance.

2.3 Corrective actions

The most important aspect of any equipment QC programme is the process that triggers some corrective action to be taken before performance deteriorates to a critical level, beyond which the equipment can no longer be used. So called '*action thresholds*' are therefore set more tightly than critical levels, as the purpose is to give an early warning of system deterioration, and thus allow time for preventative action to be taken. A detected subtle deterioration in some aspect of equipment performance (e.g. planar uniformity) may not be apparent from subjective visual assessment of the uniformity image. A major problem (e.g. spontaneous failure of a photomultiplier tube) will be obvious from the unprocessed image, but more subtle problems may not. Action thresholds are thus generally based on derived quantitative measurements, making the QC management process simple and allowing key performance parameters to be graphed on a trend line as described in Section 9.6.2.

If a camera fails its quality control because a fault has developed, it is clearly important to provide the manufacturer with as much relevant information as possible in order for a provisional diagnosis to be made and for the engineer to order possible replacement parts. The role of remote diagnostics (by using a secure virtual private network (VPN) connection to run proprietary software not normally accessible to the user) is very useful in this regard and this has become an essential feature of modern manufacturers' service support.

Apart from image quality problems, many other types of gamma camera system problem may occur (e.g. mechanical faults or gantry-computer communication errors) that are not designed to be detected by planned QC testing, and these can arise without warning. These problems should also be recorded centrally, in a camera logbook or separate 'camera faults' spreadsheet stored on a shared network drive. It is impossible to be prescriptive as to the required action in individual circumstances, but a decision clearly needs to be made (and recorded) for each and every problem as to whether to call the manufacturer immediately or to undertake some relevant first-line testing locally first. The course of action will depend on the local expertise available, but all nuclear medicine departments should have access to a Medical Physics Expert (MPE) to provide advice, even if that person is based off-site.

Both the 'camera QC' spreadsheet and the 'camera faults' spreadsheet or logbook should contain reference to other relevant documents, service reports and emails where appropriate. When it proves necessary to call out the service engineer, these two documents provide a means of retrospectively checking the performance of the manufacturer with regard to the terms of the maintenance contract (e.g. did they meet a commitment to '24 h call out').

For any type of detected fault (arising from routine QC or not), it is especially important for individual problems to be 'signed off' on completion by a responsible officer, usually a nuclear medicine physicist with MPE status. Indeed, a check of any outstanding issues arising from previous problems (i.e. inspection of both the camera QC and camera faults record) should be part of the monthly or quarterly QC procedure.

2.4 Frequency of testing

The frequency with which QC tests are performed should be set locally, but this report includes guidance on recommended intervals for each test (Section 9.6.1). The required frequency is ultimately determined by the likelihood of certain types of problems (electrical, mechanical, computer, etc.) occurring. As technology improves and equipment becomes more reliable, it is natural to expect the required frequency of some tests to decrease, but this has to be proven by local experience.

Time spent on camera QC testing generally impinges on clinical imaging time and this will always be an issue in busy departments. There is thus pressure to reduce the time spent on equipment QC, and a sensible balance has to be struck. Manufacturers are now beginning to offer an option for fully automatic QC testing whereby uniformity measurements (taking 30 minutes to several hours to perform) can be programmed to take place out of working hours (i.e. during the night) without operator intervention, thus avoiding any conflict with clinical work. QC results still have to be checked and authorised before clinical imaging starts for the next day, but this is a 5-minute task. The number and range of complex system measurements that can be performed unattended in this way are clearly expected to increase, but some (e.g. those involving a ^{57}Co sheet source) will probably remain manual.

2.5 Radiation protection considerations

Many of the QC measurements described in this report require the preparation and handling of high-activity sources. When handling liquid sources, a protective laboratory coat or plastic apron and disposable gloves should be worn. Care must be taken to avoid the transfer of possible surface contamination to equipment. This is particularly important when performing intrinsic measurements with the collimator removed. Radiation shielding should be used where possible, and sources should be kept shielded when not in use. Staff should maintain an adequate distance between themselves and the sources while measurements are in progress.

2.6 Associated computers

A gamma camera is normally interfaced to a dedicated nuclear medicine computer. The performance of the computer is unlikely to change with time, although it may also suffer acute failures or become increasingly unreliable. The performance of the computer nevertheless affects the overall quality of a nuclear medicine study; for example, it may introduce additional dead-time losses. The computer should therefore be considered to be an integral part of the system, and checks should be made to monitor its performance.

The software package provided by the computer supplier will normally include a wide range of data processing and clinical application software. Testing of manufacturers' software is considered in Chapter 8. Any faults in software should be investigated, and faults in commercial software should be reported to the supplier.

2.7 Quality assurance programme

An equipment QC programme should be viewed in the wider context of quality assurance, which should cover all of the key areas of the department's activities. This wider context is documented by a quality management system, and some nuclear medicine departments have gained certification to ISO 9001, which is the relevant international standard (ISO, 2008). Inspection of equipment QC documentation (procedures and records) forms an essential part of the periodic re-certification process.

A properly documented gamma camera QC programme is an essential requirement for a nuclear medicine department, and this is embedded in various pieces of legislation, standards, and guidance documents, most notably:

- QC is a legal requirement in the UK under regulation 32 of The Ionising Radiation Regulations 1999 (IRR, 1999), which requires employers to have a suitable quality assurance programme in place for their radiation equipment.
- In the UK, for an individual clinician to operate as a practitioner under the Ionising Radiation (Medical Exposure) Regulations (IR(ME)R) (IRMER, 2000), the Medicines (Administration of Radioactive Substances) (MARS) regulations (MARS, 1978) require that the doctor obtains a certificate from the Administration of Radioactive Substances Advisory Committee (ARSAC). The ARSAC application form has a section (appendix C) that requires a local medical physics expert to sign to confirm that regular QC is performed on various pieces of critical equipment, including gamma cameras and associated computer systems (ARSAC, 2010).
- Equipment QC is an emerging requirement for participation in national clinical trials involving nuclear medicine imaging.
- The British Nuclear Medicine Society (BNMS) policy statement on *Standards of Delivery of a Nuclear Medicine Service* states that "performance of equipment, in particular gamma cameras and dose calibrators, shall be assessed at suitable intervals" and that "records shall be maintained and be available on request" (BNMS, undated). The method of compliance with this standard is detailed in the associated organisation audit document (BNMS, 2013).
- QA/QC requirements will undoubtedly be part of developing departmental accreditation schemes (EANM, undated) and are usually required for participation in national audits of individual clinical tests.

3 Assessment of planar performance

Richard Lawson

3.1 Introduction

The simplest imaging mode for a gamma camera is when it is used to take individual images of specific parts of a patient or a phantom. This is known as planar imaging because it produces a simple projection of the source in a single plane. Planar images may be acquired as static, dynamic or cardiac gated series, but the performance requirements of the camera are the same for all of these. All gamma cameras (apart from a few specialised single photon emission computed tomography (SPECT) systems) can acquire planar images, and so assessment of planar performance forms the starting point for all gamma camera quality control. This chapter will describe the planar tests that should be carried out as part of acceptance testing of a new camera and again at suitable intervals during its lifetime. Chapters 4, 5 and 6 will discuss further tests that are required for cameras that are used for whole-body imaging and for SPECT. The tests described in these chapters are suitable for conventional gamma cameras that use a single large scintillation crystal coupled to multiple photomultiplier tubes. Chapter 7 will discuss how these tests may be modified to deal with some of the latest gamma cameras that use different technology.

In order to appreciate how gamma camera performance can be affected by the acquisition conditions used, it is necessary to have an understanding of how the gamma camera works. In particular, it is important to be aware of how the energy window is set and how corrections such as energy correction, linearity correction, and sensitivity correction (sometimes called uniformity correction) are applied. Although details of these features vary between manufacturers, interested readers will find more details in a recent IPEM publication 'The Gamma Camera – A Comprehensive Guide' (Lawson, 2013). Before performing any of the tests described in this chapter, users should understand the basics of how a gamma camera works and be familiar with the corrections that are available on the particular model of camera to be tested.

3.2 General test conditions

The majority of clinical investigations in nuclear medicine are performed with the radionuclide ^{99m}Tc, and it is therefore not surprising that many quality control (QC) procedures should be carried out using this radionuclide. Where the short 6-h half-life of ^{99m}Tc is a problem, for example, when it necessitates frequent refilling of a phantom, ^{57}Co may be an acceptable alternative because it has a long half-life of 9 months. The principal gamma ray energies emitted by ^{99m}Tc (140 keV) and ^{57}Co (122 keV) are similar, but it should not be assumed that ^{57}Co will necessarily produce the same results as ^{99m}Tc. A properly set up gamma camera should produce similar performance at 122 keV and 140 keV, but a badly designed or incorrectly adjusted gamma camera could produce different results at each energy. Moreover, ^{57}Co also emits an additional high-energy gamma ray and sources may also contain other contaminants with high-energy emissions (see Section 3.3.2). The differences between ^{57}Co and ^{99m}Tc can therefore become significant, particularly for uniformity measurement (Section 3.3). If the camera is to be used for other radionuclides as well as ^{99m}Tc, then it is important that acceptance test measurements, in particular of uniformity, are made at the energies of all radionuclides to be used, although measurements using ^{99m}Tc (or ^{57}Co) will be adequate for daily tests.

The choice of energy window used in the pulse-height analyser can also have a significant effect on the measured value of the various performance parameters. Once again, the best choice is that used for routine clinical imaging. With many gamma cameras, the recommended energy window is 10% either side of the photopeak energy, making a 20% window width (126–154 keV for ^{99m}Tc and 110–134 keV for ^{57}Co). However, with many modern gamma cameras, a window width of 15% is now recommended by the manufacturer (130–150 keV for ^{99m}Tc and 113–131 keV for ^{57}Co). Before performing any of the tests listed here, it is important to check that the energy window is correctly set and centred on the photopeak energy.

Many manufacturers specify regular tuning procedures for their gamma cameras to set the detectors to their optimal operating conditions. Before performing any of the QC tests in this chapter, particularly uniformity, the manufacturer's recommended daily and weekly tuning procedures should be carried out. However, no additional special adjustments, such as acquiring new corrections, should be performed. The aim is to ensure that the camera is working as it would for routine patient studies.

Count rate will also affect some measurements, particularly of uniformity. Owing to the limited time available for these measurements, there is a temptation to use as high a count rate as possible. However, if the count rate is too high, an erroneous result may be obtained. The maximum count rate that can be used will depend upon the dead-time characteristics of the camera (Section 3.6). All gamma cameras should be able to cope with a count rate of 20 000 counts per second (cps), although many modern cameras can handle count rates up to 40 000 cps without distortion. The count rate should generally be kept below 20 000 cps unless the manufacturer's specification explicitly allows higher rates to be used for certain tests.

Some QC measurements are made without any collimator fitted to the gamma camera. These are called *intrinsic measurements* because they reflect the performance of the detector itself. Intrinsic measurements tend to be the ones favoured by the manufacturer. However, clinical studies are always acquired with a collimator fitted and QC measurements made with the collimator are called *system (*or *extrinsic) measurements.* System measurements are more relevant to routine clinical testing and so they are the ones favoured by the user. System measurements should be made with each collimator that is to be used clinically. Most tests assume that a parallel hole collimator is used. If collimators with other hole geometries (such as diverging, fan-beam and pin-hole collimators) are used, then spatial resolution and sensitivity will vary across the field of view, and this must be taken into account when measurements are made.

The performance of a gamma camera may be expected to deteriorate somewhat towards the edge of the field of view. NEMA defines the *useful field of view* (UFOV) as the actual field of view used for imaging – with the collimator fitted (NEMA, 2012). For measurements made with the collimator removed, a lead mask should be used to limit the field to the UFOV. The *central field of view* (CFOV) is defined to be the same shape as the UFOV but with 75% of the linear dimensions of the UFOV, i.e. three-quarters of the width and three-quarters of the height. This excludes the edges of the image where performance may be worse.

With a modern digital gamma camera, performance can only be assessed using the digital images available from its associated computer. When making quantitative measurements from these images, the choice of pixel size may have a significant effect on accuracy. The optimum pixel size will depend on the test being performed and this will be discussed for each individual test.

3.3 Uniformity

For an ideal gamma camera, the image of a uniform distribution of radioactivity (sometimes called a flood source) should show the same count density everywhere across the image. In practice, the image will show areas of increased and decreased count density. These non-uniformities reflect regional changes in the performance of the camera, caused mainly by spatial non-linearity and variations in energy response. Modern gamma cameras include correction circuits to compensate for these image imperfections. The set-up of the corrections is quite sensitive to proper adjustment of the detector. If the camera is performing correctly, the final image should be uniform. Conversely, if anything goes wrong with the camera, the image uniformity is likely to deteriorate. Analysis of the uniformity of a flood image at regular intervals provides a simple test of the stability of camera performance. Any change in measured uniformity is a sensitive, albeit non-specific, indication of a change in performance which will need further investigation. Uniformity measurements should always be made with the manufacturer's corrections for energy, linearity, and sensitivity applied as would be the case for clinical imaging. However, it can sometimes be instructive to measure camera uniformity with sensitivity correction (sometimes called uniformity correction) turned off, because sensitivity correction could be used to mask an inherently poor camera performance.

3.3.1 Sources for intrinsic uniformity measurement

Intrinsic uniformity measurements are carried out with the collimator removed but replaced by a mask which limits the field of view to the same size. Whenever working with the uncollimated detector, great care must be taken not to let anything hit the crystal as it is extremely fragile. It is also important to avoid any rapid changes in room temperature as this can lead to a cracked crystal. The temperature should not be allowed to change by more than 3 °C per hour whilst the collimator is removed. When the collimators are removed, some models of gamma camera require dummy collimators consisting of thin plastic sheets to be fitted instead. These provide some degree of protection for the crystal, but care must still be maintained. In addition, removal of the collimators usually also removes the protection provided by collision sensors on the collimator face, so the gantry must then be moved with extra caution.

Without the collimator, the only way to generate a uniform activity distribution is to place a small volume source some distance away from the detector face. If the source is sufficiently far away, then the flux of gamma rays incident on the crystal is approximately uniform. The source must be positioned on the central axis of the detector and at a distance of at least 5.5 times the maximum dimension of the crystal (i.e. for a rectangular crystal, 5.5 times the diagonal). Many texts suggest that a distance of five times the field of view is adequate, but if the photon flux is to vary by less than 1% from one part of the crystal to another, then a distance of 5.5 times the diagonal is required (Lawson, 2013).

When performing any test with the uncollimated gamma camera, it is important to ensure that there are no other sources of activity nearby that could affect the measurement. This is particularly difficult in a department with more than one camera, because when uniformity measurements are being made on one camera, there may be patients containing activity in adjacent rooms. The shielding offered by intervening walls and doors should be checked and additional lead screens installed if necessary. If the camera being assessed can be turned to face completely away from all patients, this can help, but it may not be sufficient if radiation scattered from the floor, ceiling or walls can still reach the detector. Scattered radiation can be a problem if there are large activities, including injected patients, nearby because the camera has to process many extra low-energy events. Even though these fall outside the accepted energy window, if there are a large number of such events, the extra processing time can affect the camera's ability to deal correctly with good events. Once the collimator is removed and the detector is in the required position, it is therefore essential to check that the background count rate is low and to examine the background energy spectrum before placing the source to be used.

If the source can be placed far away from the detector, then an activity of about 25 MBq ^{99m}Tc will give a count rate of about 20 000 cps, but the exact activity can be adjusted depending on the distance and the permitted count rate for the gamma camera. Because the source is far away, its volume is not important and a volume of up to 1 mL in a sealed test tube or a syringe with the needle replaced by a blind cap is adequate. It is important to ensure that the line of sight from the source to the detector is not interrupted by any piece of equipment or by any person walking between the two. In addition, it is necessary to prevent small-angle scatter from the source reaching the crystal because this can lead to a non-uniform gamma ray flux. A small lead shield round the back and sides of the source can be helpful to prevent gamma rays from hitting the floor or walls and then scattering into the detector.

One way to check that the source is correctly positioned on the central axis of the camera is to leave the collimator on and move the source until its blurred image appears central in the field of view. Then the position of the source is marked and the detector angle noted. The collimator can then be removed and the detector returned to the same angle.

A more accurate method is to take a small flat mirror and mark a cross near its centre with black pen. Then, using sticky tape, this can be attached to the collimator face so that the cross is positioned exactly in the centre of the field of view (FOV). If the collimator is covered with a pressure-sensitive pad which may not be exactly parallel to the collimator surface, it is better to remove the collimator and very carefully attach the mirror to the face of the detector instead. Then a small dental inspection mirror is held near to where the expected source position will be. The user stands with their back to the camera and looks into the dental mirror as illustrated in Figure 3.1. They should adjust the angle of the dental mirror until they can see a reflection of the camera with the flat mirror attached. Then they simply move the position of the dental mirror around until the reflection of the dental mirror itself appears positioned over the cross on the flat mirror attached to the camera. In this position, the dental mirror must be on a line perpendicular to the centre of the camera face. This can then be marked as a suitable position for the source and the corresponding detector angle noted so that this can be reproduced when required.

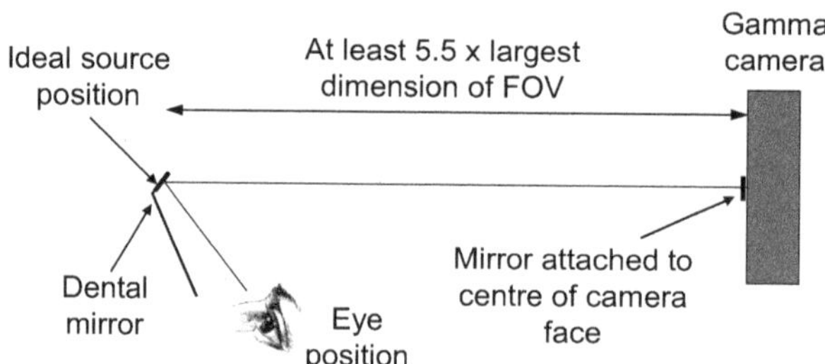

Figure 3.1 Illustrating a method of finding the ideal source position for intrinsic flood acquisition

If it is not possible to place the source sufficiently far away from the detector, then it is necessary to correct for the variations in the incident flux across the detector face. The count rate will vary across the crystal face due to the fact that the gamma ray flux falls off as the inverse square of distance and the edges are further from the source than the centre. In addition, the oblique angle of incidence near the edge spreads the gamma rays over a larger area. These two effects together produce a count rate that falls off as the cube of the cosine of the angle of incidence (Lawson, 2013). To correct for this, a curvature correction must be applied. The magnitude of the curvature correction is critically dependent on knowing the exact position of the source. This usually requires that a source holder supplied by the manufacturer is utilised and that the manufacturer's software is used to apply the correction. If the source is close to the detector then only a small activity is needed and the volume should be very small to make it point-like.

3.3.2 Sources for system uniformity measurement

Uniformity measurements made with the collimator in place require the use of a *flood source*. This may be either a solid source, consisting of a long-lived radionuclide such as ^{57}Co embedded in plastic, or a liquid-filled phantom that can be filled with ^{99m}Tc when required. Both types of flood source need to be handled carefully because of radiation protection considerations. The surface dose rate from these sources can be very high and so handling procedures must be in place to minimise operator dose, particularly to the fingers. When not in use, flood sources should be stored in a low-occupancy room or within a lead-lined container. For ^{57}Co and ^{99m}Tc flood sources, a suitable container may be constructed using 2-mm-thick lead, sandwiched between sheets of plywood.

Since the flood source is to be used for testing the uniformity of the gamma camera, it is clearly important that the flood source is itself uniform. For quantitative measurements, the uniformity of a flood source should correspond to a coefficient of variation (CoV) of better than 1% and a maximum variation from the mean of less than ±2%. This can be tested by using a scintillation detector connected to a scaler-timer to acquire counts from sample areas over the source. A lead mask with a 1-cm diameter hole should be placed over the scintillator to act as a collimator and the detector must be kept at exactly the same distance from the flood source for each measurement. At least 100 000 counts should be acquired at each position in order to give 0.3% accuracy to each measurement.

An alternative way to check the uniformity of a flood source is to acquire a system uniformity image with the flood source and then divide this by an intrinsic uniformity image. The ratio will show the combined effect of collimator non-uniformity and source non-uniformity. If this ratio image appears to be non-uniform, then the possibility that this is due to the collimator can be checked by rotating the source through 180° and repeating the measurement. If the non-uniformity pattern changes, then it may be attributed to the source rather than the collimator.

a) Solid flood sources

Solid ^{57}Co flood sources are commercially available from several suppliers. They are convenient because they are ready to use and light to handle. They can even be viewed from two sides at once. For a double headed camera, the flood source can be placed between the two heads and both acquired simultaneously. The half-life of ^{57}Co is about 9 months and so, as the source gets older, its activity will decay to the point where the count rate is too low to acquire sufficient counts in an acceptable time. Therefore, the source will need replacing at least every 18 months and the cost of regular replacement must be budgeted for. There will also be a disposal cost associated with getting rid of the old source. Sometimes this can be included in the cost of purchasing a replacement source on an exchange basis. If old sources are kept to decay, then the registration with the appropriate national regulatory agency must be sufficient to enable the user to hold several sources at once.

Flood sources of ^{57}Co usually contain small amounts of impurity, mainly ^{56}Co and ^{58}Co and possibly also ^{60}Co. These isotopes emit higher energy gamma rays, of 800 keV and above, which can significantly affect the performance of the camera. This is particularly noticeable when measurements are made with a low-energy, high-resolution (LEHR) collimator. This has a low efficiency for 122 keV gamma rays (because it collimates them correctly), but hardly reduces the intensity of high-energy gamma rays (because they easily penetrate the collimator septa). Thus, even a small percentage of the high-energy impurities can have a noticeable effect on camera performance. The count rate in the 122 keV energy window is hardly affected, but the high count rate outside the window leads to increased dead time and pulse pile up in the electronics. In some cameras, this can significantly degrade image uniformity. However, even if impurities are kept low, ^{57}Co itself also has 0.2% of emissions at 690 keV so the problem can never be completely avoided. Since the effect relates to the absolute count rate produced by the high-energy emissions, this is only a real problem with newly purchased, high activity (e.g. 750 MBq) sources (Busemann Sokole et al., 1996). Therefore, caution should be exercised before purchasing sources with too high an activity, and the images acquired with new sources should be compared with those from an older source before the new source is used routinely. Since ^{56}Co and ^{58}Co both have half-lives of less than 12 weeks, the magnitude of the effect decreases with time, but it is clearly not economic to buy a new source and not be able to use it for a few weeks. The problem can be minimised by placing the source some distance (e.g. 50 cm) away from the collimator instead of directly on it. This reduces the number of high-energy gamma rays that reach the crystal, because of the inverse square law, without changing the number of 122 keV gamma rays detected, since collimator efficiency is independent of distance.

b) Liquid filled flood sources

Refillable liquid sources can be filled with ^{99m}Tc solution whenever required and so they avoid the problems of high-energy emissions from ^{57}Co and also save on the replacement costs. However, they are not easy to make or to use.

One possible design for a liquid-filled source is shown in Figure 3.2. The 20 mm gap between the two Perspex faces is filled with water to which ^{99m}Tc solution can be added. In order to achieve a uniform source, it is vital that the two Perspex faces are perfectly flat and exactly parallel with a maximum deviation of $\pm 2\%$, which means a 0.2 mm tolerance on each sheet. Therefore, the Perspex sheets used to construct the source must be of uniform thickness and undistorted after construction. In addition, in order to ensure that the faces do not bow under the weight of the water inside, they must be quite thick – at least 20 mm – which makes the sources quite heavy. However, the extra thickness is also an advantage since it increases the amount of scattering material in the source, which makes the energy spectrum received by the camera more realistic, so the camera performance should be more similar to that achieved in patient studies.

To fill the phantom, both filler screws are removed and water is added until the phantom is nearly full, but leaving a good sized air bubble. Then the required

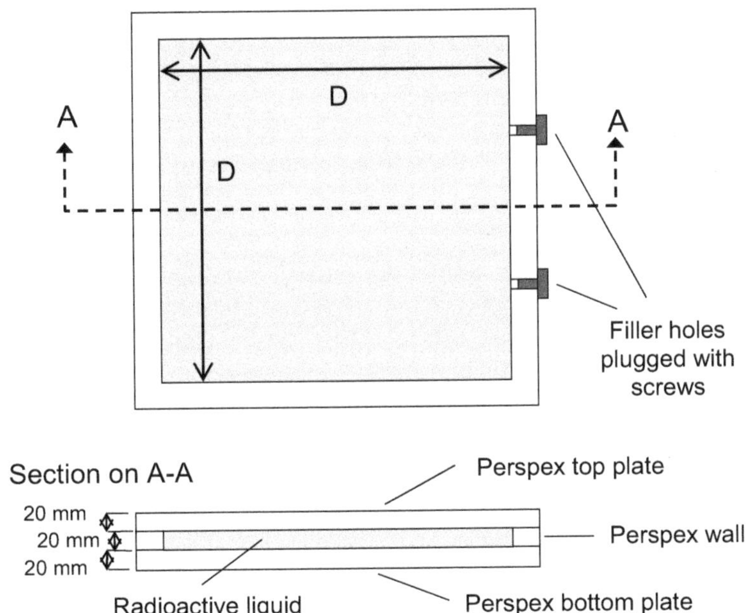

Figure 3.2 Design of a liquid filled flood source. The dimension 'D' must be greater than the field of view of the gamma camera

activity is added and the filler screws are replaced. Some radiopharmaceuticals can stick to the plastic, so it is safest to just use ^{99m}Tc in the form of pertechnetate. The activity can then be mixed thoroughly by tilting the phantom back and forth so that the air bubble swishes around through all corners of the phantom several times. Because of the weight of the phantom and the radiation protection issues of handling it for too long, this is not straightforward to achieve by hand. It can be made easier by mounting the phantom on a suitable spindle mount. In order to achieve adequate mixing, the air bubble must be large enough to allow good circulation of the water, but once the activity is thoroughly mixed, more water can be added to reduce the size of the bubble. As long as the active area of the source is larger than the field of view of the gamma camera, any remaining air bubble can be positioned outside the camera's field of view during the uniformity measurements by tilting the phantom slightly. If the source does not need to be used immediately, then, after a quick initial mix, the uniformity of the activity distribution can be improved by leaving it to stand for a few hours.

Manual mixing becomes impossible for large sources which are too heavy to handle. However, commercial versions are available containing a pump to facilitate mixing. This avoids the difficulty of having to mix the source manually, but it can take several minutes of pumping to achieve adequate mixing before the source is ready to use.

If a phantom of this type is proven to be flat and uniform when new, care must be taken to ensure that it remains so during use. The pressure of water inside together with ageing of the materials can cause a phantom to become distorted over time. Therefore, when a new phantom is constructed or purchased, it is a good idea to place a metal straight-edge across each face to check that it is flat within the required 0.2 mm and to repeat this check from time to time during routine use.

3.3.3 Uniformity image acquisition

Because uniformity depends upon the energy of the radionuclide, routine measurements should be made with the radionuclide most commonly used with the camera, which in most cases will be ^{99m}Tc. Before clinical tests are carried out with a different radionuclide, the uniformity at that energy should also be measured.

Some manufacturers allow the system to store a range of uniformity correction maps, using a different one for each collimator. In order to assess how good clinical images may be, it is reasonable to apply the appropriate uniformity correction for the collimator to be used. However, in order to see whether camera performance is changing with time, it is a good idea to also acquire an image with no uniformity correction applied or with a fixed uniformity map applied. A new correction map should only be acquired when the current one is shown to be inadequate. If a new correction map is always acquired before uniformity is quantified, then any changes will remain masked.

For uniformity measurement, there is no point in using an image pixel size which is less than the distance over which the camera response might change. A 128 × 128 matrix gives a pixel size of around 3 to 4 mm for most sizes of camera. This is just equal to the intrinsic spatial resolution and with a conventional single crystal camera, the uniformity is unlikely to change significantly in that distance. Therefore, for most cameras, a 64 × 64 matrix is optimum since this gives a pixel size of around 6 to 8 mm which is greater than the spatial resolution, and changes from one pixel to the next could represent genuine non-uniformities. For quantitative analysis, in order to achieve 1% statistical accuracy, at least 10 000 counts per pixel are required. A 64 × 64 matrix has 4096 pixels, but circular or rectangular cameras do not completely fill the square matrix so only about 3000 pixels are used. This means that at least 30 million counts must be acquired in total. It is counterproductive to use a 128 × 128 matrix because four times as many counts would be needed and there is no need to know how uniformity changes at a scale of only 3 to 4 mm. However, data may be acquired on a 128 × 128 (or 256 × 256) matrix and then converted to 64 × 64 before analysis, provided that the conversion is done by adding counts in groups of four (or 16) pixels.

Some manufacturers' software packages are known to perform matrix size conversion in odd ways that are not suitable for uniformity analysis. If just one pixel value out of each group of four is used, then noise is increased, even though this value is multiplied by four. If the four values are averaged rather than added, then the noise no longer follows a Poisson distribution (where the standard deviation is the square root of the pixel count) and so any corrections for noise (Section 3.3.4) would need to be modified. Images with fewer than 10 000 counts per pixel will be dominated by random statistical noise and so only large changes in camera uniformity will be detectable. Therefore, if the count rate is low and time only allows shorter acquisitions with fewer than 10 000 counts per pixel, then uniformity quantification will generally be unreliable and uniformity analysis should be limited to a visual assessment unless a correction is applied (Section 3.3.4).

3.3.4 Uniformity image analysis

Uniformity can be assessed either by visual examination of the flood image or by quantitative measurements. Visual inspection is useful as a quick assessment that there are no gross abnormalities in the image and this may be acceptable for a routine daily check before any patients are imaged. However, for a more detailed assessment of uniformity, provided adequate counts are acquired, some sort of quantitative analysis is required. This is particularly useful in detecting slow changes in uniformity which allows planned intervention before they become too severe.

Quantitative measures of uniformity can be divided into two types: those which measure global changes in uniformity and those most effective for detecting local effects. A quality control protocol should contain measures of each type. NEMA recommends using *integral uniformity* as a global measure and *differential uniformity* as a local measure (NEMA, 2012). However, this report also recommends *coefficient of variation* as a global measure and suggests that *spread of differential uniformity* can serve as an alternative local measure. Manufacturers of nuclear medicine computer systems will usually provide programs that calculate the NEMA parameters of integral and differential uniformity. However, if the computer system includes tools that calculate the minimum, maximum, mean and standard deviation of counts in a region of interest, then it is easy to use these figures to calculate the integral uniformity and coefficient of variation using the formulae shown below. Calculation of the spread of differential uniformity is more complex and this will require access to user written programs. These uniformity values should be calculated for both the UFOV and CFOV.

a) Integral uniformity

Integral uniformity is simply a measure of the contrast difference between the hottest and coldest pixels anywhere in the flood image. NEMA specify that the image should have at least 10 000 counts per pixel and that it must be smoothed using a nine-point filter before calculations are performed (NEMA, 2012). This is necessary to reduce the largest fluctuations which are due to random noise. If C_{max} represents the counts in the hottest pixel and C_{min} the counts in the coldest pixel in the field of view, then NEMA define

$$\text{Integral Uniformity} = \frac{C_{max} - C_{min}}{C_{max} + C_{min}} \times 100\% \tag{3.1}$$

A large value for integral uniformity implies a poor uniformity and a small value good uniformity. For a modern gamma camera, intrinsic integral uniformity is typically between 2% and 3% in the CFOV and slightly worse over the UFOV, although this depends on the uniformity corrections that have been applied to the image. Integral uniformity has the advantage that it can usually be calculated without the need for programming as long as the image has sufficient counts and it is smoothed with the correct filter first. However, it can be criticised as a measure of global uniformity because it only uses the information from two out of the several thousand pixels in the image and it gives no indication of how bad other pixels might be.

b) Coefficient of variation

The coefficient of variation gives a measure of the overall variation in pixel counts by taking account of all pixels across the field of view. If $C(i,j)$ represents the counts in the pixel at row i and column j of the flood image and there are N pixels in the field of view, then the mean pixel count, M, is given by

$$M = \frac{\sum_{i,j} C(i,j)}{N} \tag{3.2}$$

The standard deviation of all pixels in the field of view, SD, is given by

$$SD = \sqrt{\frac{\sum_{i,j} (C(i,j) - M)^2}{N}} \tag{3.3}$$

The coefficient of variation, CoV, is defined as the standard deviation of the pixel counts expressed as a percentage of the mean pixel count.

$$\text{CoV} = \frac{SD}{M} \times 100\% \tag{3.4}$$

If all of the individual pixel counts were plotted as a histogram showing how often each value occurred, then the CoV would represent the width of the distribution showing how far pixels deviate from the mean value on average. A small CoV implies that all pixel values are very close to the mean and hence the image has good uniformity. Note that the sum over i and j in Equation 3.2 and Equation 3.3 is performed over all pixels in the UFOV or CFOV as appropriate and N is the number of pixels included in the sum. Unlike the NEMA integral uniformity, the image does not require smoothing first. The main limitation of using CoV as a measure of uniformity is that it is influenced by noise in the image and so the value obtained depends on the counts acquired. Therefore, it is not possible to quote a typical value for CoV without also specifying the average counts per pixel.

c) Correction for Poisson noise

The variation in pixel counts in a flood image arises from two sources: genuine camera non-uniformity and random statistical fluctuations, or noise. Since the noise is caused by random independent radioactive decays, it will follow a Poisson distribution with a standard deviation equal to the mean. Therefore, if there are 10 000 counts per pixel, the noise component will be 1% and this should enable real non-uniformity to be detected as long as it is greater than 1%. However, if the non-uniformity is small or there are insufficient counts in the image, then the coefficient of variation can be dominated by random noise and it will not represent a true figure for the non-uniformity. It is easy to perform a correction for this, as first suggested by Cox and Diffey (1976). The corrected variance (square of the corrected SD) is equal to the measured variance (square of the measured SD) minus the variance due to the Poisson noise. Since the variance due to Poisson noise will just be M, this gives for the corrected standard deviation:

$$\text{Corrected SD} = \sqrt{SD^2 - M} \quad (3.5)$$

and so the corrected coefficient of variation will be:

$$\text{Corrected CoV} = \frac{\sqrt{SD^2 - M}}{M} \times 100\% \quad (3.6)$$

The corrected CoV will be a better representation of the true uniformity than the CoV. It has the advantage that it does not depend on the number of counts acquired in the image. A modern gamma camera can achieve an intrinsic corrected CoV of less than 1% over the CFOV. In order to detect small non-uniformities, it is still necessary to acquire adequate counts, but by applying this correction to the coefficient of variation, it is possible to obtain a valid measure of camera uniformity with fewer than the 30 million counts that are normally recommended (Lawson, 2013).

A study using simulated artefacts added to gamma camera uniformity images has shown that daily QC images can be acquired with as few as 5 million counts but, although the corrected CoV gave a more accurate value of the non-uniformity than the CoV, it was in fact no better at detecting typical non-uniformity patterns, provided that the change in value from baseline was used (Murray et al., 2014).

d) Differential uniformity

NEMA suggests a measure of differential uniformity based on how rapidly counts change in a local group of five pixels. If H is the highest count in a group of five adjacent pixels in a row or column and L the lowest count in the group, then NEMA define a quantity D (NEMA, 2012):

$$D = \frac{H - L}{H + L} \times 100\% \quad (3.7)$$

This is a measure of the contrast between maximum and minimum counts in this small area of the image. D is calculated for every possible group of five adjacent pixels in all rows of the image and also for all possible groups of five adjacent pixels in all columns of the image. This gives thousands of values for D and NEMA then define the differential uniformity to be the largest value of D obtained anywhere in the image. In other words, it is the highest local contrast anywhere in the image. Once again, this measure can be criticised as it shows only the worst value in the image, and does not indicate how uniform the rest of the image might be. As for integral uniformity, the image should have at least 10 000 counts per pixel and should first be smoothed with a nine-point filter. A typical value of differential uniformity for a modern gamma camera is between 1.5% and 2.5% in the CFOV and slightly worse over the UFOV.

e) Spread of differential uniformity

While the coefficient of variation is a good measure of global variations in uniformity, it does not say anything about how rapidly the uniformity varies. For example, a large coefficient of variation could be due to rapid changes in uniformity over a small area, or the same changes spread gradually from one side of the image to another. The NEMA differential uniformity does look at how quickly counts change in a local area but it only chooses the single worst contrast value from Equation 3.7. The spread of differential uniformity is a measure that overcomes this limitation by including contrast values from all groups of five pixels and so it gives a number that reflects the full range of contrast changes in the image. It was first proposed as a useful index of camera uniformity by Hughes and Sharp (1989) and they recommend it as being able to detect small localised non-uniformities as well as being sensitive to global degradation in non-uniformity. It can be a helpful parameter to calculate if suitable software is available.

For each pixel in the image, the contrast between it and the five following pixels on the same row are calculated.

$$D_{ijh} = \frac{C(i,j) - C(i+h,j)}{C(i,j) + C(i+h,j)} \quad \text{for } h=1 \text{ to } 5 \tag{3.8}$$

Similarly, the contrast between this pixel and the five following pixels in the same column are calculated.

$$D_{ijk} = \frac{C(i,j) - C(i,j+k)}{C(i,j) + C(i,j+k)} \quad \text{for } k=1 \text{ to } 5 \tag{3.9}$$

That gives 10 values of D for every pixel and the calculation is repeated for every pixel in the field of view to give several thousand D values. Then the spread of differential uniformity is defined as the standard deviation of all of the D values. Note that since the D values are equally likely to be positive or negative, the mean D value will be zero. Therefore

$$\text{Spread of differential uniformity} = \sqrt{\frac{\sum_{ijh} (D_{ijh})^2 + \sum_{ijk} (D_{ijk})^2}{N}} \tag{3.10}$$

where N is the total number of D values calculated.

Although Equation 3.10 is expressed in a different way to that used in the previous IPEM report (IPEM, 2003a), its meaning and value remain the same. If all of the D values were plotted as a histogram showing how frequently each value occurred, then the mean of the histogram would be zero and the spread of differential uniformity would represent the width of the histogram. In contrast, the NEMA differential uniformity would be just the most extreme value in the histogram.

3.3.5 Practical uniformity measurement

The recommended protocol for quantitative assessment of gamma camera uniformity is very similar for both intrinsic and system measurements.

Protocol for uniformity measurement

a) Perform any daily or weekly tuning procedures recommended by the manufacturer.

b) For an intrinsic measurement, remove the collimator and place a small volume source of ^{99m}Tc at a distance of at least 5.5 times the size of the diagonal of the field of view from the detector.

c) For a system measurement, fit the required collimator and place a flood source of ^{99m}Tc or ^{57}Co in contact with the collimator face or close to it. For dual head systems, it may be possible to place a suitable source midway between the two detectors so that they can both be acquired simultaneously. Four upturned plastic cups on the lower collimator may be used to support a suitable ^{57}Co flood source midway between the two detectors.

d) The activity should be sufficient to give a count rate as high as possible without introducing distortions due to pulse pile-up. Do not exceed the maximum count rate recommended by the manufacturer, or 20 000 cps if no other limit is specified.

e) Set the appropriate energy window for the radionuclide used. Check that the energy peak is central in the window. Turn on relevant uniformity corrections as used for clinical imaging.

f) Acquire an image on a 64×64 matrix with at least 30 million counts, or at least 10 million if a Poisson correction is to be applied.

g) If the image has at least 10 000 counts per pixel, smooth the image with a nine-point filter and calculate the NEMA integral and differential uniformity.

h) Using the unsmoothed image, calculate the coefficient of variation of the image pixels and apply the Poisson correction.

i) If possible, calculate the spread of differential uniformity.

j) If it was not possible to acquire them simultaneously, repeat the measurements for each detector.

k) Plot the calculated values on a graph for comparison with those obtained in previous weeks.

l) Determine whether any action thresholds have been reached and, if so, take appropriate remedial action.

3.3.6 Offset windows test

A gamma camera is normally operated with an energy window set around the photopeak energy of the nuclide being imaged. Under these conditions, the uniformity of the image should be optimum, because this is how the camera is designed to operate. If the energy window is set slightly above or below the photopeak, then the image will become less uniform. However, by deliberately offsetting the energy window, some faults in the camera may be emphasised and can become more obvious. Therefore, a pair of intrinsic uniformity images acquired with offset windows can be an aid to diagnosis; a 'low window' image is acquired with the centre of the energy window set 10% below the photopeak and another 'high window' image with the centre of the energy window set 10% above the photopeak. Because energy, linearity, and sensitivity corrections are only designed to work properly with a photopeak energy window, it is usually necessary to turn all these corrections off when performing this test. Under these conditions, both the low window and the high window image will appear very non-uniform and the pattern of the photomultiplier tubes will be very obvious. However, some camera faults can be identified by comparing the two images.

For example, if one photomultiplier tube is giving a lower signal than the others, then it will appear relatively 'hot' compared with other tubes on the low window image and relatively 'cold' on the high window image. Conversely, a photomultiplier tube that is giving a higher than average signal will appear relatively cold in the low window image and relatively hot in the high window image.

The offset windows test can also be useful for detecting hydration of the crystal caused by moisture getting in. The sodium iodide crystal assembly is manufactured in a dry environment and then encapsulated to prevent moisture from the air entering. If the encapsulation breaks down, then moisture may get in and this causes the crystal to turn yellow in places. The yellowing blocks some of the light transmission and so reduces the size of the energy signal locally. Therefore, an irregular area at the edge of the crystal, which appears relatively hot in the low window image, may be a sign of yellowing caused by breakdown of the encapsulation.

A fault during manufacture of the crystal can allow moisture in at positions far away from the edge. These may cause small areas of yellow crystal which show up as small spots that are relatively hot in the low window image and relatively cold in the high window. This measles like appearance is quite characteristic of hydration because nothing else produces such small well defined artefacts.

The International Atomic Energy Agency quality control atlas shows a nice example of these effects (IAEA, 2003).

3.3.7 Frequency of uniformity measurement

Uniformity should be measured as part of the acceptance testing of any new gamma camera. This must be done using NEMA recommended procedures (NEMA, 2012) to verify that the camera is performing to the manufacturer's specifications. The NEMA uniformity test will usually be carried out by the installation engineer, but it should be witnessed by the user in order that they can become familiar with the manufacturer's procedure and verify the result. At the same time, the user should carry out a quantitative assessment of uniformity using the protocol recommended in this chapter. This should be performed for both intrinsic and system measurements using all available collimators. The results of one of these (either intrinsic or system uniformity with a specified collimator) will form the baseline for a regular weekly measurement of uniformity.

A uniformity image should be acquired with the gamma camera each day before the first patient is imaged. This is necessary to ensure that the camera is functioning correctly and that there have been no component failures since the camera was last used. However, it is not necessary to quantify the uniformity every day. For daily QC, a flood image with at least 5 million counts should be acquired and inspected visually for any serious non-uniformity. This daily QC may be performed either as an intrinsic measurement (with the collimator removed) or a system measurement (with the collimator in place) but the schedule should incorporate some system measurements so that each available collimator can be used from time to time to check for unsuspected collimator damage. If a collimator is suspected of having been damaged during use, it should be subject to a full quantitative assessment of uniformity before being used again.

At least once a week, a full quantitative assessment of uniformity should be made as described in the above protocol. This will require more counts to be acquired in order that one or more of the above parameters can be measured. The results should be compared with those obtained in previous weeks and action taken if any action levels are exceeded (Section 9.6.2). A full quantitative assessment of uniformity should also be performed after the camera has been serviced or repaired, before it is put back into clinical use.

The offset windows test does not need to form part of regular QC but it may be helpful if the results of routine uniformity testing appear inexplicably poor.

3.4 Spatial resolution

The spatial resolution of a gamma camera is a measure of the sharpness of the image produced. It is usually measured by the width of the profile through the image of a point or line source. If two points or two lines are closer together than this distance, they merge into one and cannot be resolved, whereas if they are further apart, they can be resolved as two separate objects. Spatial resolution is not a measure of the size of the smallest feature which will be visible in an image. Provided that the contrast is sufficiently high, even isolated features smaller than the camera's resolution may be visualised, although they will appear larger than they really are and with reduced contrast.

Two approaches can be taken to assess spatial resolution: qualitative and quantitative. For qualitative measures, a test object is imaged by the camera and the resolution determined by the amount of detail that is visible in the image. This has the advantage of involving the whole imaging process, including the display and observer. For quantitative measures, alterations in detail size produced by the imaging process are objectively measured.

There are two components to the spatial resolution of the gamma camera. The *intrinsic resolution* represents the reproducibility of the calculated coordinates for gamma rays incident at the same site in the crystal. Note that this does not reflect the accuracy with which the coordinates have been calculated, as this also depends upon the spatial linearity of the camera. The *system resolution* also includes the effect of the collimator which changes significantly with distance. Intrinsic resolution and collimator resolution combine together in quadrature (in the same way that the sides of a right-angled triangle combine to give the hypotenuse) to give the system resolution.

$$(\text{System Resolution})^2 = (\text{Intrinsic Resolution})^2 + (\text{Collimator Resolution})^2 \qquad (3.11)$$

There are also two types of source that can be used to measure resolution: either a line source or a point source. This report describes methods for measurement of both intrinsic and system resolution using either line or point sources.

3.4.1 Qualitative assessment of resolution

Spatial resolution can be assessed qualitatively by imaging a suitable test phantom and many such phantoms have been proposed. One type uses the minimum detectable size of some artificial lesion in an anatomical structure, such as the thyroid phantom or liver slice phantom. These may be useful for demonstrating certain features of gamma camera operation, but they do not form part of the quality control process.

A second group consists of transmission phantoms made of lead sheet with holes or slots cut to allow gamma rays through. By irradiating the transmission pattern with a uniform source, a high contrast image is cast onto the camera detector. For intrinsic measurements, the collimator is removed and the transmission phantom is placed against the crystal in its place. Then a small volume source at a distance (as used for intrinsic uniformity – Section 3.3.1) can be used to irradiate it. For system measurements, the transmission phantom is placed against the collimator and a uniform flood source placed immediately behind it. However, with some phantoms this can lead to Moiré interference patterns between the regular array of holes in the phantom and holes in the collimator (Hart, 1986). This effect can just be seen in the two bottom left segments of the phantom in Figure 3.3. The effect is only likely to be a problem when measurements are being made at the highest resolution, i.e. at the collimator face. Depending on the design of the collimator, this effect may be reduced by rotating the test pattern to an alternative position relative to the collimator.

Two commonly used patterns are the Anger pie phantom (Figure 3.3) and the quadrant bar phantom (Figure 3.4). The minimum resolvable bar spacing or hole size is related to the spatial resolution, although it is not equal to it. In Figure 3.3, the segment containing the 1.25 mm holes separated by 5 mm is just resolved, so the resolution is often said to be 5 mm, although this is only an approximate estimate and not a formal measurement. The results are not unduly dependent upon image count density. Either phantom can be used for intrinsic measurements, but the Anger phantom is less useful for system resolution because the fraction of gamma rays transmitted through its holes is rather small, so when used with a collimator, a high activity flood source is required.

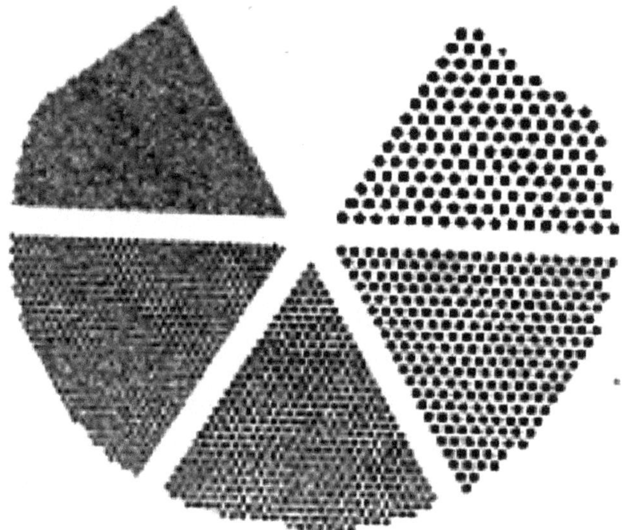

Figure 3.3 Qualitative resolution image obtained with an Anger pie phantom. The five visible segments have holes with diameter 1.0, 1.25, 1.5, 2.0, and 2.5 mm separated by a distance of four times the diameter. The 1.25 mm holes separated by 5 mm are just resolved so the resolution is about 5 mm

Figure 3.4 Qualitative resolution image obtained with a quadrant bar phantom

Such patterns provide a quick visual assessment of resolution but do not give a proper quantitative measure. The number of sets of bars (or holes) limits the accuracy of the method, and the results are dependent upon the effectiveness of the display and the visual performance of the observer. However, as long as conditions are reproducible, such images taken at regular intervals can be used to check that performance has not deteriorated over time.

The minimum resolvable bar spacing of the quadrant bar phantom is not the same as the width of the image of a line source that is used as a formal quantitative measure of resolution (Section 3.4.2). However, by examining the intensity modulation of any set of visible bars, it is possible to deduce the equivalent width of the line spread function (Wasserman, 1998).

Protocol for qualitative assessment of resolution

a) For an intrinsic measurement, remove the collimator and carefully place the transmission phantom in contact with the detector. Place a small volume source (less than 1 mL) containing a few hundred megabecquerels of ^{99m}Tc at a large distance from the collimator.

b) For a system measurement, fit the required collimator. Place the transmission phantom in contact with the collimator. Place a flood source of ^{99m}Tc or ^{57}Co on top of the phantom.

c) Set the appropriate energy window for the radionuclide used. Check that the energy peak is central in the window.

d) Acquire an image containing 1 million counts using a pixel size of about 1 mm (i.e. at least a 512×512 matrix).

e) Note the spacing of the smallest bars or holes that can be resolved in the image and relate this to actual spacing from the known dimensions of the phantom.

f) Where the design of the phantom allows, note whether the minimum resolvable feature also changes towards the edges of the field of view.

g) For a system measurement, repeat with the transmission phantom positioned 5 cm and 10 cm away from the collimator face.

h) Repeat for each detector and for a system measurement with all available collimators.

3.4.2 Quantitative measurement of resolution

The image of a line or point source of radioactivity gives a direct measurement of the degree of image blurring. Provided that the width of the line or point is much smaller than the camera's resolution, its width on the image represents the spatial resolution of the camera. A profile taken through the image of a line yields the *line spread function* (LSF) and a profile through the image of a point gives the *point spread function* (PSF). Note that the LSF and PSF are different as will be demonstrated below.

The full width at half maximum (FWHM) of the LSF or PSF is commonly used as a measure of spatial resolution (Figure 3.5a). The FWHM is approximately equal to the minimum separation required between two line sources if they are to be just resolved. (This is analogous to the Rayleigh criterion in optics, although, in that

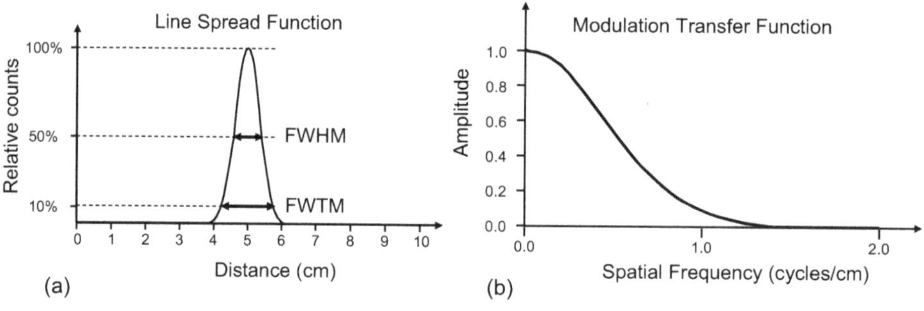

Figure 3.5 Quantification of resolution. a) Line spread function and b) modulation transfer function

case, the profile is caused by a diffraction pattern which is not the case here.) However, the FWHM only gives partial information about resolution and so it is sometimes supplemented by the full width at tenth maximum (FWTM) height of the LSF or PSF (Figure 3.5a) which gives additional information about the wings of the profile. The FWTM can be particularly useful for system resolution where it can be affected by septal penetration.

To define adequately the shape of the LSF or PSF, the acquisition pixel size should be less than or equal to one-fifth of the FWHM (NEMA, 2012) which equates to about 0.7 mm for measurement of intrinsic resolution with most gamma cameras. For a large gamma camera with a 500 mm field of view, this means that a 1024 × 1024 matrix must be used, giving a pixel size of 0.5 mm. If the computer system does not allow a 1024 × 1024 matrix, then a 512 × 512 matrix with a zoom factor of two can be used instead (although only one-quarter of the field of view will be imaged at once). If the pixel size is too large, there may be fewer than five pixels across the profile and there is a risk of missing the true peak value. This is only likely to be a significant problem with intrinsic resolution because system resolution is much larger.

For statistical precision, the PSF or LSF should have at least 10 000 counts in the peak point. This means that, if a line source is used, it is only necessary to have about 200 counts in the highest pixel of the image, because a broad profile can be taken that sums 50 adjacent rows and still gives 10 000 counts in the centre of the profile. However, if this is done, care must be taken that the line is positioned exactly perpendicular to the direction of the profile. This needs to be done quite carefully otherwise the profile will be artificially broadened. If the centre of the line must not deviate by more than 1 pixel across the width of the profile then, in order to be able to take a 25 mm wide profile (summing 50 rows), the line needs to be aligned with the X or Y axis to within 1 degree. Profiles should not be summed over a width of more than 25 mm because this puts unreasonable demands on the accuracy of alignment. In addition, non-linearities in the image appear over distances comparable with the size of photomultiplier tubes (PMTs), and so summing over greater widths can cause profiles to become broadened by non-linearity rather than just resolution. Even if linearity correction is applied, intrinsic resolution varies between positions at the centre of a PMT and midway between PMTs and this difference will be averaged out if profiles are broader than 25 mm.

If a point source is used, 2000 counts in the maximum pixel of the image should give about 10 000 counts maximum when a broad profile is taken. With a point source, there are no problems of alignment and both horizontal and vertical profiles can be taken through the same source to give both X and Y resolution from the same image. As shown below, a broad profile taken through a point source gives the LSF rather than the PSF. The profile just needs to be wide enough to encompass the tails of the distribution.

When a profile has been obtained, the peak height can be determined and the full width at half maximum (FWHM) and full width at tenth maximum (FWTM) can then be measured in pixels. It will probably be necessary to use linear interpolation to find the exact points, in fractions of a pixel, corresponding to 50% and 10% of the peak height on each side. The problem is more difficult if the pixel size is too large, because the highest pixel value may be an underestimate of the true peak height. In this case, it will be necessary to fit a Gaussian curve to the peak to determine the true peak height. Gibson (1992) has suggested a method of dealing with profiles that are only coarsely digitised.

The FWHM and FWTM can then be converted from pixels to millimetres using a calibration of mm/pixel obtained from the distance between two lines. The X and Y resolution should be measured at a range of positions across the field of view and the average X resolution and average Y resolution should be quoted. However, it is known that intrinsic resolution will be worst at positions close to the centre of each PMT and best for positions between PMTs (Lawson, 2013). Therefore, if the PMT positions can be identified, it is also instructive to determine the average intrinsic resolution at the centre of PMTs compared with the average resolution at positions between PMTs. A modern gamma camera will have an intrinsic spatial resolution with FWHM between 3.0 mm and 4.0 mm averaged over the CFOV, with very similar figures over the UFOV. System resolution is mainly dependent on collimator resolution and so values obtained are greatly influenced by collimator design. It is easy to design a collimator with good resolution at the expense of poor sensitivity and *vice versa*, so system resolution for any collimator cannot be judged in isolation without also considering its sensitivity (Nuclear Fields, 2013).

a) Relationship between line spread function and point spread function

Figure 3.6a shows part of an infinitely long line source with unit activity per unit length. By definition, a narrow profile drawn across the line gives the LSF. The counts in the profile at a distance x from the line are due to activity from all points

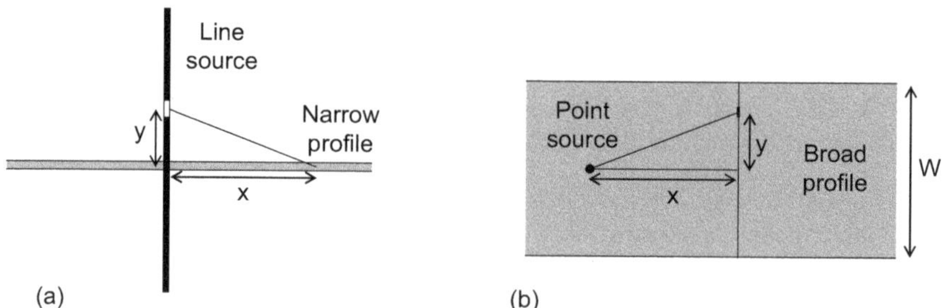

Figure 3.6 Illustrating the relationship between line spread function and point spread function. a) The geometry of a narrow profile taken across a line source. b) The geometry of a broad profile taken across a point source

along the line. The contribution from the bit of the line at position y is given by the PSF at a distance $\sqrt{x^2+y^2}$. Therefore, when contributions from the full length of the line are integrated, we have

$$\mathrm{LSF}(x) = \int_{-\infty}^{\infty} \mathrm{PSF}\left(\sqrt{x^2+y^2}\right) dy \tag{3.12}$$

This demonstrates that the LSF and PSF are different. The LSF will always be broader than the PSF because of the inclusion of components from more distant parts of the line.

Since the profile is the same everywhere along the line, we get the same result for the LSF whether the profile is narrow or wide. In practice, the line does not have to be infinitely long, provided that the profile does not come too close to the end of the line.

In contrast, if we use a point source, we get different results for narrow and wide profiles. By definition, a narrow profile gives the PSF, but a wide profile involves a summation of several profiles. Figure 3.6b shows a point source of unit activity and a broad profile of width W including the source at its centre. The counts in the profile at a distance x are given by a summation of many PSFs

$$\text{Wide Profile}(x) = \int_{-W/2}^{W/2} \mathrm{PSF}\left(\sqrt{x^2+y^2}\right) dy \tag{3.13}$$

Comparing Equation 3.12 and Equation 3.13 shows that, as long as W is large enough such that the PSF has fallen to a negligible value at $\pm W/2$, a wide profile through a point source gives the LSF, just the same as a profile through a long line. This result means that it is sometimes convenient to acquire an image of a point source but then by creating a broad profile, it is the line spread function that is determined.

b) Modulation transfer function

The FWHM and FWTM are just two numbers that describe something about the width of the LSF, but they do not describe its full shape. One could of course plot the full curve of the LSF as in Figure 3.5a, but it is more useful to plot the modulation transfer function (MTF) as in Figure 3.5b. The MTF shows how the gamma camera responds to a wide range of spatial frequencies. If the camera is used to image a set of parallel hot and cold bars with a wide spacing (i.e. a small number of lines per cm), then the bars will be clearly resolved and the contrast between hot and cold is preserved. This corresponds to an amplitude of 1.0 at low spatial frequencies in Figure 3.5b. If the spacing of the hot and cold bars is progressively reduced (increasing number of lines per cm), then the poor resolution will reduce the contrast between hot and cold areas. This corresponds to the falling amplitude of the MTF with increasing spatial frequency in Figure 3.5b. Two methods of calculating the MTF at four spatial frequencies from an image of the quadrant bar phantom have been described by Wasserman (1998).

In fact, the MTF is just the Fourier transform of the LSF (Lawson, 2013). Therefore, provided the LSF is symmetrical, the MTF can be calculated using the following equation:

$$MTF(u) = \frac{\int_{-\infty}^{\infty} LSF(x)Cos(2\pi ux)dx}{\int_{-\infty}^{\infty} LSF(x)dx} \tag{3.14}$$

where $MTF(u)$ represents the modulation transfer function at spatial frequency u and $LSF(x)$ represents the value of the line spread function at position x. A program for this calculation has been published by Benedetto and Nusynowitz (1977) with a modification suggested by Neff et al. (1977). The MTF can be applied to the quantitative measurement of both intrinsic and system resolution.

3.4.3 Intrinsic resolution using a line source

The precautions for working with an uncollimated detector already described in Section 3.3.1 should be observed. With the collimator removed, a transmission slit phantom is placed across the detector face. The slits should be no wider than 1 mm. Unless the field of view is circular or square, it is necessary to have two separate phantoms with the slits in orthogonal directions: one for X and one for Y resolution measurement. Manufacturers usually have phantoms of this type specifically designed to fit their own cameras but they are not always available for use by users. A source is placed at a large distance from the detector to give a uniform illumination of the slits as already described for intrinsic uniformity measurement.

Protocol for intrinsic resolution using a line source

a) Remove the collimator and place the slit phantom so that it runs parallel to the X axis of the acquisition matrix. It is important that the phantom is aligned to better than 1°.

b) Position a source of ^{99m}Tc at a distance of about five times the detector diameter.

c) Set the energy window for ^{99m}Tc. Check that the energy peak is central in the window.

d) Set up a computer acquisition using a pixel size no larger than 0.7 mm.

e) Acquire an image for a time sufficient to give about 200 counts in the maximum pixel.

f) Process the acquired image by drawing several profiles across the lines in the Y direction. Each profile should be no wider than 25 mm.

g) Determine the image pixel size using the number of pixels between the peaks of the first and last lines in the profiles and measuring the actual spacing of the corresponding slits in the phantom.

h) Measure the FWHM and FWTM of each profile where it crosses each line. Convert the values from pixels to millimetres using the calculated pixel size.

i) Calculate the average Y resolution (both FWHM and FWTM) from all lines in all profiles.

j) Calculate the average Y resolution from all lines that coincide with the centre of PMT positions (i.e. where the width of the lines is greatest).

k) Calculate the average Y resolution from all lines that fall midway between PMTs (i.e. where the width of the lines is smallest).

l) Turn the slit phantom through 90 degrees or replace it with one where the lines run parallel to the Y axis of the acquisition matrix. Repeat the measurements to determine the X resolution.

m) Repeat the measurement for each detector.

3.4.4 Intrinsic resolution using a point source

In the absence of the collimator, it is necessary to collimate the source itself to provide a collimated beam of gamma rays. Figure 3.7 shows how a suitable

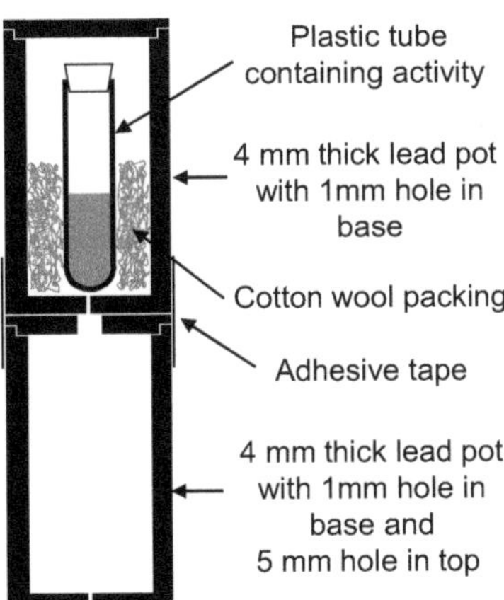

Figure 3.7 Construction of a collimated point source from two lead pots

collimated point source can be constructed from lead pots. Two lead pots each have 1 mm diameter holes drilled in the centre of their base. The two pots are then stacked on top of each other and held together by binding with adhesive tape. A small plastic tube containing some ^{99m}Tc solution can then be placed in the top pot so that a fine pencil beam of gamma rays emerges through the bottom hole. It is best to put the activity in a small sealed plastic tube that can be held centrally in the upper lead pot using cotton wool packing so that as much activity as possible is centralised over the holes. Even so, it will require a few hundred megabecquerels of activity to give an adequate count rate through the small hole. Therefore, the lead pots need to be at least 4 mm thick to minimise background to other parts of the crystal. If normal 2 mm thick lead pots are used, it is necessary to place them inside an outer pot in order to provide enough shielding.

The precautions for working with an uncollimated detector already described in Section 3.3.1 should be observed. If the collimator is removed and the detector positioned facing upwards, then the collimated point source needs to be placed as close to the detector as possible with the beam projecting downwards, but the source must not be placed directly on the detector because of the risk of damaging the crystal. If the camera has dummy collimators that are fitted when the collimator is removed, then the source may be placed on the dummy. The high degree of collimation incorporated in the source shown in Figure 3.7 ensures that the gamma ray beam will still be narrow when it reaches the crystal, even though the dummy is 2 cm away from the detector surface. If no dummy collimators are provided, then the crystal should be protected using a 5 mm thick sheet of foam modelling board. This can be purchased from craft suppliers and is easily cut to size to fit across the face of the detector. The source may then be placed gently on the modelling board.

The source can be placed at selected positions in the field of view to measure resolution at positions that are centred over the middle of a PMT and other positions that are midway between PMTs. If several such sources are available, then time can be saved by measuring the resolution at several positions at once. Both X and Y resolution can be measured from the same acquisition by taking profiles in the required direction.

Protocol for intrinsic resolution using a point source

a) Remove the collimator and protect the detector with a dummy collimator or a sheet of foam modelling board. Mark the positions of each PMT.

b) Position a few collimated point sources over the centre of selected PMTs.

c) Set the energy window for ^{99m}Tc. Check that the energy peak is central in the window.

d) Set up a computer acquisition using a pixel size no larger than 0.7 mm.

e) Acquire an image for a time sufficient to give about 2000 counts in the maximum pixel.

f) Process the acquired image by drawing a wide profile across each source in the Y direction. Measure the FWHM and FWTM of each profile. Convert the values from pixels to millimetres using the image pixel size to determine the Y resolution. Pixel size must be determined in a separate experiment.

g) In a similar manner, draw wide profiles across each source in the X direction to determine the X resolution.

h) If necessary, acquire more images with the sources placed over the centre of different PMTs until about 12 tubes have been measured. Calculate the average X and Y resolution at the centre of PMTs.

i) Repeat with the sources placed midway between PMTs.

j) Repeat the measurement for each detector.

3.4.5 System resolution using a line source

A suitable line source can be made from a glass capillary or polythene tubing of internal diameter not greater than 1 mm. With polythene tubing, care must be taken to ensure that the line source remains straight, either by taping it to thick card or placing it in a groove machined in a thin sheet of Perspex. The tube must be filled fairly uniformly with ^{99m}Tc solution without introducing any air bubbles. If two or more lines are available, then several lines can be acquired at once.

Measurements should be made with the line source in direct contact with the collimator surface and repeated at distances of 5 cm and 10 cm from the collimator face. The acquired pixel size should be less than one-fifth of the expected FWHM, but since system resolution is worse than intrinsic resolution, this is not too difficult to achieve.

Protocol for system resolution using a line source

a) Fit the required collimator.

b) Place the line source in contact with the collimator so that it runs across the centre of the field of view. Align it accurately with the X axis of the acquisition matrix. It is important that the phantom is aligned to better than 1°.

c) If a second line is available, place it parallel to the first line and exactly 100 mm from it.

d) Set the energy window for ^{99m}Tc. Check that the energy peak is central in the window.

e) Set up a computer acquisition using a pixel size no larger than 1 mm.

f) Acquire an image for a time sufficient to give about 400 counts in the maximum pixel.

g) If only one line is available, move it by exactly 100 mm and acquire another image.

h) Process the acquired images by drawing several profiles across both lines in the Y direction. Each profile should be no wider than 25 mm.

i) Determine the image pixel size using the number of pixels between the peaks of the two lines in the profiles and the known distance of 100 mm between them.

j) Measure the FWHM and FWTM of each profile. Convert the values from pixels to millimetres using the calculated pixel size.

k) Calculate the average Y resolution (both FWHM and FWTM) from all lines in all profiles.

l) Turn the line through 90° so that it runs parallel to the Y axis of the acquisition matrix. Repeat the measurements to determine the X resolution.

m) Repeat with the line positioned at distances of 5 cm and 10 cm from the collimator face. It is not necessary to repeat the pixel size measurement at each distance.

n) Repeat the measurement for each detector and for all available collimators.

3.4.6 System resolution using a point source

Since there is a collimator fitted to the camera, there is no need to collimate the source for this measurement. A suitable point source may be made by placing a small drop of ^{99m}Tc, with as high a radioactive concentration as possible, into a small hole in a sheet of Perspex. A 1 mm diameter hole drilled to a depth of 2 mm in a 5 mm thick sheet of Perspex is suitable. A Perspex sheet with several holes drilled allows several positions to be measured simultaneously. A syringe with a fine needle and a steady hand are required to make sure that the activity goes into the hole without spreading around it. All of the holes should have approximately the same activity. After filling, the holes should be sealed with adhesive tape to avoid spillage.

Protocol for system resolution using a point source

a) Fit the required collimator.

b) Place the point phantom in contact with the collimator so that the point sources are distributed across the field of view.

c) Set the energy window for ^{99m}Tc. Check that the energy peak is central in the window.

d) Set up a computer acquisition using a pixel size no larger than 1 mm.

e) Acquire an image for a time sufficient to give about 2000 counts in the maximum pixel.

f) Process the acquired image by drawing a wide profile across each source in the Y direction.

g) Determine the image pixel size using the number of pixels between the peaks of two profiles and measuring the actual distance between the corresponding sources.

h) Measure the FWHM and FWTM of each profile. Convert the values from pixels to millimetres using the known pixel size to determine the Y resolution.

i) Calculate the average Y resolution (both FWHM and FWTM) from all point sources.

j) In a similar manner, draw wide profiles across each source in the X direction to determine the X resolution. Determine the pixel size, convert into millimetres and calculate the average X resolution.

k) Repeat with the point phantom positioned at distances of 5 cm and 10 cm from the collimator face.

l) Repeat the measurement for each detector and for all available collimators.

3.4.7 Frequency of resolution measurement

A quantitative measurement of intrinsic resolution should be made as part of the acceptance testing of any new gamma camera. System resolution should also be measured with each available collimator. These values will form a baseline for comparison with future measurements. If qualitative phantoms are available, images may also be acquired with these to act as a reference.

Thereafter, it is only necessary to check resolution on an annual basis or if other tests (such as uniformity) show poor performance. Intrinsic resolution should be checked once a year either quantitatively or qualitatively and compared with the reference acquired when the camera was new. Collimator resolution should be checked if damage is suspected.

3.5 Spatial linearity

While spatial resolution refers to the blurring of the image, spatial linearity measures the spatial distortion. This distortion is most easily appreciated by assessing the straightness of the image of a line source. Without linearity corrections, the non-linearity can be very pronounced and easy to measure, but once linearity corrections are applied, the remaining non-linearity is small (typically 0.5 mm) and so this becomes difficult to measure. Since clinical images will always be acquired with linearity corrections applied, there should be no obvious non-linearity in the images. Since non-linearity can have a marked effect on image uniformity, any problems should become apparent when examining uniformity (Section 3.3).

Sources for assessing linearity include the slit phantom with parallel equally spaced lines as used for intrinsic resolution measurement (Section 3.4.3) and the orthogonal hole phantom with an array of small holes arranged in a regular square pattern.

Protocol for assessment of linearity

a) Place the test pattern on the uncollimated face of the camera. If using a slit phantom, make sure that the slits are exactly parallel with the Y axis of the acquisition matrix.

b) Position a source of ^{99m}Tc at a distance of at least five times the detector diameter.

c) Set up an acquisition using the smallest available pixel size.

d) Acquire an image containing 1 million counts.

e) If using a slit phantom, rotate it through 90° and obtain another image with the slits parallel to the X axis.

f) For a qualitative assessment, just examine the images for any deviations from linearity. Patterns of non-linearity will usually coincide with the positions of the PMTs. For checking non-linearity, it is the centre of the image of the line or points that matters; the width may also vary but this is due to changes in spatial resolution which also follow the pattern of the PMT positions.

g) For a quantitative assessment, draw horizontal profiles (in the X direction) across the image. For the slit phantom, the profiles should be taken at 1 cm intervals across the full Y range of the image. For the orthogonal hole phantom, the profiles should be taken at the Y position of each hole.

h) From the profiles, determine the X position of each line (or hole) at every Y position. Perform a linear least-squares fit to these values to determine the best fit to each line. Measure the differences (in pixels) between the fitted line and the imaged line. Convert these values to millimetres using the known pixel size.

i) Calculate the mean and standard deviation of these differences.

j) Note the maximum deviation between the fitted and imaged lines.

k) Repeat the analysis using vertical profiles on the other image.

l) Repeat the measurement for each detector.

3.5.1 Frequency of linearity testing

Linearity should be tested as part of the acceptance testing of a new gamma camera. Thereafter, it only needs to be checked if another test, such as uniformity, shows unexpected results. Any significant change in linearity will lead to a detectable change in uniformity.

3.6 Count rate capability

As the activity viewed by the gamma camera is increased, the count rate should ideally increase linearly with activity. However, in practice, when the count rate gets too high, the electronics tend to miss some events and so the observed count rate does not rise as rapidly as expected. Figure 3.8 shows a typical count rate capability graph where the observed count rate is plotted against expected count rate – which should be proportional to source activity. At low count rates, the observed and expected count rates are the same and the graph follows a straight line corresponding to the true rate. But as the expected count rate (or activity) increases,

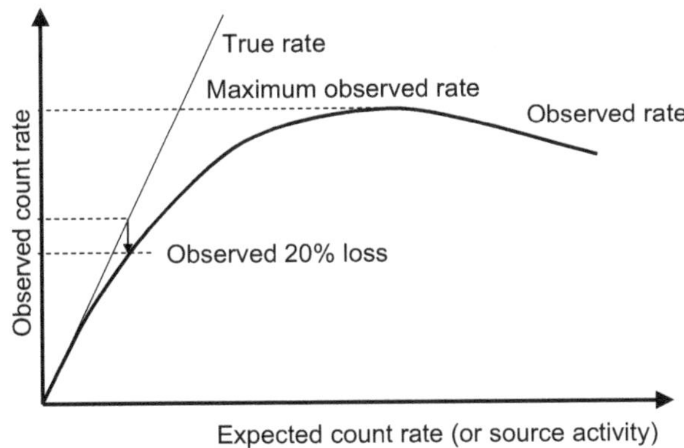

Figure 3.8 Typical curve illustrating count rate performance of a gamma camera

the graph becomes non-linear and the observed count rate falls below the true rate. At extreme count rates, the observed count rate can even fall with increasing activity.

The behaviour of gamma camera electronics can be described as being either *paralysable*, *non-paralysable* or a combination of the two (Lawson, 2013). The non-paralysable detector will not record another gamma ray until a certain time, known as the dead time, has elapsed since the previous event. In a paralysable system, the arrival of another event within this dead time will not only result in it being lost but will actually initiate a new dead-time interval during which no events will be recorded. That is why the count rate graph can actually fall with increasing activity.

At very high count rates, image quality can also deteriorate because the electronics is not able to position events correctly, but, for the majority of clinical studies, count rates are unlikely to approach the levels where there is a significant degradation in image quality. Even in those involving high activities, such as first-pass cardiac studies, it is primarily the quantitative data which suffer significant error. Measurement of count rate capability is thus primarily aimed at establishing the relationship between observed count rate and expected count rate.

Count rate performance is expressed quantitatively by quoting the observed count rate at the point where 20% of the counts are lost, i.e. where the observed count rate is 80% of the expected count rate. For a modern gamma camera, this may be in excess of 200 000 cps. The maximum count rate recorded is also often quoted although this has no clinical value as the image quality will have become unusable at this point.

Observed and expected counts refer only to events that fall within the set energy window, but the camera electronics will also have to cope with many events that have an energy outside the window and end up being rejected. If more dead time is lost in dealing with the rejected events, then more good events will also be missed. Therefore, the count rate capability is greatly influenced by the fraction of events that fall within the accepted energy window, and hence by the width of the energy window and the shape of the energy spectrum, which in turn depends on the scattering conditions. The most linear response is achieved when events are distributed uniformly across the detector and there is no scattered radiation. The least favourable is when imaging a small concentrated source situated in scattering material. Therefore, different results will be obtained for intrinsic and system measurements. However, system measurements are difficult to obtain because of the very high activity needed to push to the limits of the curve, so count rate capability is usually performed as an intrinsic measurement because much less activity is needed.

3.6.1 Intrinsic count rate performance

For an intrinsic measurement, the collimator is removed and the source is placed in air so there is no scattering material present. Therefore, the energy spectrum will contain the maximum number of photopeak events and so the count rate performance of the detector will be at its best. It should be recognised that the intrinsic count rate performance does not necessarily represent the true performance of the camera in a clinical situation. The energy window should be the same as the one used for clinical studies at low count rates and the window must not be allowed to change as the count rate changes.

Ideally, the whole field of view should be uniformly irradiated, which implies that the source should be placed some distance away, but this usually means that only one detector of a dual head system can be tested at once. However, uniform irradiation is not too critical and so it may be possible to place a source midway between the two uncollimated detectors and hence test them simultaneously. However, in this position, count rate will depend critically on distance, and so it is important that the source is not moved between successive measurements.

The precautions for working with an uncollimated detector already described in Section 3.3.1 should be observed. This is particularly important if the acquisitions are to be left to proceed unattended overnight or over a weekend. The camera room should be locked and other staff (such as cleaners) should be excluded. Since the collimators will be left off, it is important that the air conditioning system maintains a constant room temperature through the night, otherwise crystal damage could occur.

Protocol for intrinsic count rate performance measurement

a) Remove the collimator.

b) Set the energy window for ^{99m}Tc using a source that gives a count rate of less than 20 000 cps.

c) Make up a source of ^{99m}Tc with sufficient activity to saturate the count rate when it is placed about 1 metre away from the detector.

d) Acquire a static 5 min image using a 128×128 matrix and note the starting time of the acquisition. The 128×128 matrix is only necessary in order to avoid the counts in each pixel overflowing.

e) Repeat the static acquisition at intervals of about 1 h as the source decays making sure that the source is in exactly the same place every time. Some gamma cameras will allow this acquisition to be automated so that it can proceed unattended. This may be done through a sequence of static acquisitions or a very long dynamic acquisition. If a dynamic acquisition is used, make sure that any automatic decay correction is turned off, otherwise the activity will not seem to decrease with time.

f) Continue hourly measurements until the count rate has fallen below 1000 cps. This may require up to 10 half-lives (60 h) which implies a weekend acquisition.

g) Remove the source and obtain a final 5 min background measurement.

h) Subtract the count rate, determined from the background image, from the count rate in all of the other images to give the observed count rate for each image, $C_{obs}(i)$, where i is the image number. This is particularly important for the low count rate images that will be used to determine the expected count rate.

i) Calculate the time, $T(i)$, in minutes that has elapsed from the start of the first image acquisition to the mid-point of each subsequent image. The following steps assume that $T(i)$ is the mid-time of the image, but some computer systems may take $T(i)$ as the start time of the image and this is the convention used by NEMA (2012). In this case, a small correction may be necessary to allow for decay during the acquisition. This is done by multiplying the observed count rate by $(1+0.00096 \times \Delta t)$ where Δt is the duration of the image in minutes. This correction will be negligible if acquisitions are only 5 min long but can become significant if long acquisitions are used – for example, as part of a long slow dynamic series.

j) Apply a decay correction to the observed count rate to extrapolate back to what it would have been at the start of the first image, $C_{extrap}(i)$, using the equation $C_{extrap}(i) = C_{obs}(i) \times \exp(0.00192T(i))$.

k) Plot a graph of $C_{extrap}(i)$ against $T(i)$ for all images. For early images, the extrapolated count rate will be significantly less than the true initial count rate because of dead-time effects. However, the last few points of this graph (where the count rate is less than 5000 cps) will not be affected significantly by dead-time and so the graph should rise asymptotically towards a constant plateau. Determine the average value of this plateau and use it as the best estimate of the expected true initial count rate, $C_{exp}(0)$.

l) Calculate the expected count rate for each image, $C_{exp}(i)$, from the expected initial count rate using $C_{exp}(i) = C_{exp}(0) \times \exp(-0.00192T(i))$.

m) Plot a graph of observed count rate against expected count rate. This should look something like Figure 3.8. Plot the line of true count rate (where observed count rate equals expected count rate). Determine the observed count rate at which the observed rate is 80% of the true rate (the 20% loss point).

n) Determine the maximum observed count rate.

o) If it was not possible to acquire them simultaneously, repeat for each detector.

3.6.2 System count rate performance

System count rate performance can be measured using a distributed source incorporating scattering material. This gives a much more realistic simulation of the clinical situation because the camera electronics have to cope with the entire energy spectrum of events. There are phantoms designed specifically for this test (IPEM, 2003a; IEC, 2005; NEMA, 2012) but if these are not available, an approximation can be made using any extended source of ^{99m}Tc (at least 100 mm diameter) placed inside a container of water. Because the count rate performance depends on the distribution of activity and scattering material, the results obtained will not be exactly the same as those for the official phantom, but then every patient study is different anyway. The energy window should be fixed and the same as the one used for routine clinical studies. Because the collimator is fitted to the gamma camera, a lot of activity is needed to achieve the necessary high count rate. Measurements can be made by filling the phantom with a few gigabecquerels of ^{99m}Tc and then taking measurements as it decays as described above for intrinsic count rate performance.

Alternatively, the phantom can first be filled with a low activity and then increments of activity added. If this method is used, the activity increments need to be carefully chosen so that there are a few measurements on the low activity portion of the curve where it is still linear and some on the high activity non-linear portion of the curve. The activity added at each increment can be accurately determined by weighing the syringe from which it is dispensed, but two different concentrations will be needed: one to cover the low activity range and another to cover the high activity range. The relationship between the concentrations in the two syringes can be determined by preparing the low activity syringe from a weighed aliquot out of the high activity syringe which is then diluted by a known factor. It is important that the thickness of the source does not change during the measurements and so the volume of activity added needs to be kept small compared with the initial volume of the source.

3.6.3 Frequency of testing

If possible, count rate capability should be assessed as part of acceptance testing. Thereafter, it does not need to be tested again unless major component replacement has taken place.

3.7 System sensitivity

Sensitivity is a measure of the count rate obtained from the gamma camera for each unit of activity within the source. It is related to the proportion of gamma rays emitted from a radionuclide source which are detected within the photopeak of the collimated gamma camera. System sensitivity therefore depends on both the detection efficiency of the crystal and the geometric efficiency of the collimator. System sensitivity can range from 50 cps/MBq for an ultra-high resolution collimator to 500 cps/MBq for a high sensitivity collimator.

For a parallel hole collimator, the sensitivity is independent of the distance of the source from the face of the collimator. Thus, any source configuration should give the same value for sensitivity, but in practice, two factors need to be taken into consideration. The first factor is the possibility of attenuation of the gamma radiation, primarily by the source itself. To minimise this effect, the physical thickness of the source of radioactivity and the wall thickness of the container should be kept to a minimum. A Perspex container with walls no more than 3 mm thick and containing no more than 3 mm thickness of solution is recommended.

The second factor to be considered is that, although the response of the collimator should not depend upon the position of the source, non-uniformity variations may occur. To minimise this effect, the area of the detector irradiated by the gamma rays should be as large as possible. The container should provide an active area at least 150 mm in diameter.

Protocol for sensitivity measurement

a) Fit the required collimator.

b) Draw up an activity of a few tens of megabecquerels of ^{99m}Tc in a syringe and measure the activity in a radionuclide calibrator. Note the time of measurement.

c) Fill the chosen container with ^{99m}Tc solution from the syringe.

d) Recount the used syringe in the radionuclide calibrator to determine any residual activity remaining. Note the time of measurement.

e) Place the source on the surface of the collimator.

f) Set the energy window for ^{99m}Tc. Check that the energy peak is central in the window.

g) Acquire an image for 2 min. Note the time of the acquisition.

h) Remove the source and acquire a background image for 2 min.

i) Determine the total count rate (cps) in the source image and in the background image. Subtract the background to give the background corrected count rate in the source.

j) Decay correct the initial syringe activity and the residual activity to the time of the source acquisition. Subtract the decay corrected residual activity from the decay corrected syringe activity to give the activity in the source at the time of acquisition.

k) Divide the background corrected count rate (cps) by the decay corrected activity (MBq) to give the sensitivity (cps/MBq).

l) Repeat the measurement for each detector and each available collimator.

m) The measurement may also be repeated with other radionuclides using the appropriate collimators.

3.7.1 Frequency of sensitivity measurement

System sensitivity must be measured for each available collimator as part of acceptance testing for all new gamma cameras. Thereafter, it should be repeated at monthly intervals.

3.8 Energy resolution

Energy resolution is a measure of the ability of the detector to discriminate between different gamma ray energies. It is determined by looking at the photopeak in the energy pulse height spectrum (Figure 3.9). Energy resolution is measured by the full width at half maximum (FWHM) of the photopeak expressed as a percentage of the photopeak energy. To avoid scatter, such a measurement should be made using a small source of activity in air as for an intrinsic uniformity measurement (Section 3.3.1). The width of the photopeak measured using a flood source depends not only on the performance of the scintillation detector but also on local variations in the position of the photopeak. Modern gamma cameras incorporate corrections for these local variations within the energy corrections so, in practice, the result using a flood source is not very different. For a gamma camera using a sodium iodide scintillation crystal, energy resolution is usually around 10%.

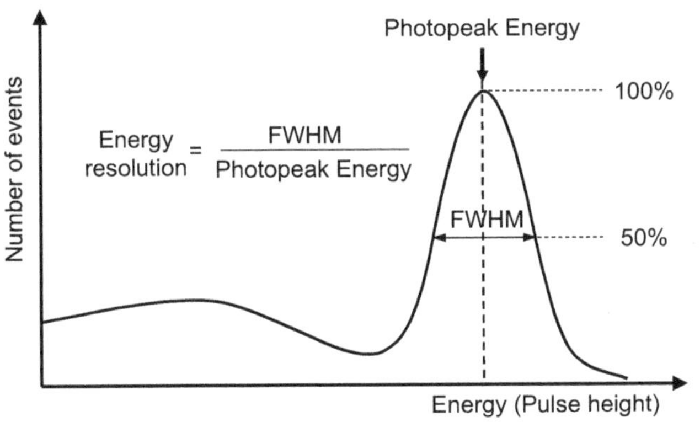

Figure 3.9 Energy spectrum illustrating calculation of energy resolution

Protocol for energy resolution measurement

a) Remove the collimator.

b) Place a small source of ^{99m}Tc at sufficient distance to give an approximate uniform irradiation of the detector. The count rate should not exceed 20 000 cps.

c) Display the energy spectrum of events on the gamma camera. Wait until the spectrum has at least 10 000 events in the peak channel and then record the ^{99m}Tc spectrum. Some gamma cameras include a utility to store the spectrum values as numbers that can be saved for analysis in a spreadsheet. Others allow hard copy of the spectrum to be printed out. If neither option is available, it should be possible to set a very narrow energy window (about 1 keV wide) and move this in small steps through the photopeak noting the count rate within the window each time, which enables the photopeak shape to be plotted by hand.

d) Without altering the gamma camera settings, collect another spectrum using a small source of ^{57}Co instead of ^{99m}Tc. Record the ^{57}Co spectrum in the same manner.

e) Examine the ^{99m}Tc spectrum and determine the position corresponding to the centre of the photopeak and the positions on either side where the counts have fallen to half the peak value. Depending on how the spectrum is recorded, these positions may be measured in terms of channel number, displayed kiloelectron volts or even distance in millimetres on a printout. Whichever method is used, call these display units.

f) Examine the ^{57}Co spectrum and determine the position corresponding to the centre of the photopeak in the chosen display units. Take care to ensure that the scale used is the same as for the ^{99m}Tc spectrum.

g) Determine an appropriate scale factor relating display units to actual energy in kiloelectron volts.

$$\text{Scale factor} = \frac{18.4}{\text{Display units between } ^{57}\text{Co and } ^{99m}\text{Tc photopeaks}} \tag{3.15}$$

where 18.4 keV is the difference between the gamma ray energies of ^{57}Co (122.1 keV) and ^{99m}Tc (140.5 keV).

h) Measure the FWHM of the ^{99m}Tc photopeak in display units from the difference between the half count values on either side of the peak.

i) Determine the energy resolution, expressed as a percentage of the ^{99m}Tc photopeak energy using

$$\text{Energy resolution} = \frac{\text{FWHM in display units} \times \text{Scale factor}}{140.5} \times 100\% \qquad (3.16)$$

j) Repeat the measurement for each detector.

3.8.1 Frequency of energy resolution measurement

Energy resolution must be measured as part of acceptance testing for all new gamma cameras. Thereafter, it should be repeated annually.

3.9 Multiple window spatial registration

Multiple window spatial registration refers to how well the gamma camera image of an object retains its size and shape when imaged with different energy windows. It can be measured by examining the difference in the detected positions of gamma rays of two different energies emanating from the same point. In old gamma cameras, image size could change significantly with gamma ray energy, but in modern cameras, normalisation circuits should overcome this gross effect. However, image non-linearities can be energy-dependent and so it is still possible to find local variations in position which vary with energy. Good multiple window spatial registration is an important consideration in dual radionuclide studies such as parathyroid subtraction imaging using ^{99m}Tc (140 keV) and ^{201}Tl (70 keV), and in the use of radionuclides with multiple energies such as ^{67}Ga (93, 185 and 300 keV). The difference in registration will normally depend on the particular gamma-ray energies and so should be measured with a source of the radionuclide (or mixture of radionuclides) used in clinical studies. Typically, multiple window spatial registration is less than 1.0 mm.

Multiple window spatial registration is an intrinsic measurement and so the source used must be a collimated pencil beam. A suitable source can be made from two lead pots with holes similar to that described for intrinsic spatial resolution (Section 3.4.4) and shown in Figure 3.7. However, if ^{67}Ga is used, the lead must be thicker because of the higher maximum energy and the holes can be larger because position rather than resolution is being measured.

The precautions for working with an uncollimated detector already described in Section 3.3.1 should be observed.

Protocol for measurement of multiple window spatial registration

a) Remove the collimator and protect the detector with a dummy collimator or a sheet of foam modelling board.

b) Position the detector head horizontal and facing upwards. Place the collimated point source assembly as close as possible to the crystal and close to the centre of the field of view.

c) Set the energy windows for the appropriate radionuclide or mixture of radionuclides being used. Each energy window should be acquired into a separate image.

d) Set up an acquisition using the smallest available pixel size.

e) Acquire an image containing at least 1000 counts in the maximum pixel.

f) Move the source by a known distance along the X or Y axis and acquire further images at several different locations.

g) On each acquired image, identify the X and Y coordinates (in pixels) of the centre of the source by calculating the centroid of the count distribution in each of the acquired energy windows.

h) For each position of the source, find the maximum difference in X or Y coordinate between any of the energy windows used.

i) Convert these pixel distances into actual millimetres using the measured number of pixels between two points and the known distance that the source was moved.

j) Repeat the measurement for each detector.

3.9.1 Frequency of testing

Multiple window spatial registration should be tested as part of acceptance testing for all new gamma cameras. Thereafter, it should be repeated annually.

3.10 Computer acquisition tests

Since the computer that controls data acquisition is now an essential part of any modern gamma camera, it is necessary to include at least some tests of the computer acquisition functions. There are relatively few published standard tests in this category, although the International Atomic Energy Agency recommends some basic tests of computer timing for static and dynamic mode and for electrocardiogram (ECG)-gated acquisition (IAEA, 2009). Some possible methods of testing static, dynamic and ECG-gated acquisition are suggested in this chapter, although no specific protocols are given because these are not formally recognised tests. Whole-body acquisition tests are included in Chapter 4 and SPECT acquisition tests in Chapter 5. Tests of data processing utility software are discussed in Chapter 8.

3.10.1 Static acquisition

Digital systems will always have a maximum possible count for one pixel and when this value is reached, counts may either saturate or recycle to zero. The behaviour of the system can be tested by using a syringe containing about 10 MBq of ^{99m}Tc placed close to a low-energy, general purpose (LEGP) collimator so that the image is confined to a small area. Acquire a series of static images using a 64×64 matrix (in order to obtain maximum counts per pixel) with durations of 1 s, 4 s, 16 s, 64 s, 256 s, and 1024 s. Note the pixel count at which saturation or overflow occurs (typically 32 767 counts per pixel). Check that the total counts in each image increase linearly with acquisition time (at least until pixel overflow occurs).

The remaining static and dynamic tests should all be performed with the same source so that the count rate remains fixed. Any constant source can be used, but a uniform source (with camera set up for either system or intrinsic measurement) is convenient because it avoids any problems of pixel overflow. The count rate should be a few thousand counts per second.

Preset count termination can be tested by acquiring static images on each available matrix size using a preset count of 100 000 counts. Examine each acquired image to verify that they all contain the specified number of counts and that the recorded duration of each image was the same.

Preset time termination can be tested by acquiring static images on each available matrix size using a preset time of 60 s. Use an independent stopwatch to check that the actual acquisition time is 60 s. Examine each acquired image to verify that they all contain the same number of counts.

3.10.2 Dynamic acquisition

Use the same uniform source that was used for the static acquisition test.

Basic timing can be tested using a dynamic acquisition of ten 60 second frames. Use an independent stopwatch to check that the total acquisition lasts for 600 s. Verify that the number of counts in each frame is the same as that from the static acquisition.

Multi-phase dynamic acquisition should also be tested by performing a dynamic acquisition with as many separate phases as are allowed by the system (for example, sixty 1 second frames, followed by six 10 second frames followed by three 100 second frames). Check that the total acquisition duration is correct and verify that the count rate (cps) is the same for every frame, taking into account the poorer statistics associated with the shorter time acquisitions.

3.10.3 ECG-gated acquisition

Use the same uniform source that was used for the static acquisition test.

Electronic ECG simulators are used for testing ECG machines so it may be possible to borrow one from the medical electronics department. If a simulator is not available, electrodes can be attached to a suitable volunteer to give a real ECG trace (but without administering any radioactivity).

The most basic test involves performing an ECG-gated acquisition with a preset number of beats and a fixed R-R interval window. Check that the system rejects beats that are outside this window by varying the rate on the ECG simulator or by briefly disconnecting an ECG lead. On completion, check that the requested number of beats have been acquired. Verify that the count rates in the first three-quarters of the acquired frames (e.g. 24 frames out of 32) are the same. It is acceptable for the count rate to drop off in the last few frames because of variations in heart rate within the window. If the system allows the R-R interval window to follow a changing heart rate, this mode can be tested by slowly changing the rate from an ECG simulator.

This test with a constant source will not detect any timing delays due to the acquisition starting too late after the R-wave of the ECG. This can be tested with a beating cardiac phantom (e.g. the Vanderbilt phantom). Since the timing depends on the polarity and duration of the trigger signal from the ECG machine, this is best checked with the actual equipment used for patients. This may mean that it cannot be tested until the first patient is imaged. A timing delay will cause the ventricular activity–time curve to appear shifted so that the maximum count no longer appears in the first frame.

3.10.4 Frequency of testing

These tests of computer acquisition should be performed as part of acceptance testing of a new camera and repeated when a new version of the acquisition software is installed.

3.11 Summary of measurement protocols

For easy reference, this section contains a summary of the acquisition parameters recommended for each of the measurements described in this chapter. Full details can be found in the relevant sections.

3.11.1 Uniformity

For full details, see Section 3.3.5

Parameter	Intrinsic measurement	System measurement
Collimator	None	All available
Source	1 mL at distance of 5.5 × FOV	Uniform flood source
Radionuclide	^{99m}Tc	^{99m}Tc or ^{57}Co
Activity	20–50 MBq	~400 MBq
Count rate	<20 000 cps unless manufacturer allows more	
Pixel size	6 to 8 mm (64 × 64 matrix)	
Counts acquired	30 million (or 10 million with a Poisson correction)	
Analysis	Global: NEMA integral uniformity, Coefficient of Variation (corrected for Poisson noise) Local: NEMA differential uniformity, Spread of differential uniformity	
Frequency	Acceptance testing – intrinsic and system tests Daily – at least qualitative check Weekly – full quantitative analysis After service – full quantitative analysis	

3.11.2 Resolution (qualitative)

For full details, see Section 3.4.1

Parameter	Intrinsic measurement	System measurement
Collimator	None	All available
Transmission Phantom	Bar or Anger pie phantom placed on detector face	Bar phantom placed on collimator
Source	1 mL at large distance	Uniform flood source
Radionuclide	^{99m}Tc	^{99m}Tc or ^{57}Co
Activity	~200 MBq	~400 MBq
Count rate	<20 000 cps	
Pixel size	~1 mm (512 × 512 matrix)	
Counts acquired	1 million total	
Notes		Repeat with phantom at distances of 5 cm and 10 cm from collimator
Frequency	Acceptance testing – optional, as a reference Annually – optional, compare with reference	

3.11.3 Resolution (quantitative) using a line source

For full details of intrinsic measurement, see Section 3.4.3 and for system measurement, Section 3.4.5

<table>
<tr><th>Parameter</th><th>Intrinsic measurement</th><th>System measurement</th></tr>
<tr><td>Collimator</td><td>None</td><td>All available</td></tr>
<tr><td>Transmission phantom</td><td>Slit phantom on crystal</td><td>None</td></tr>
<tr><td>Source</td><td>1 mL at distance of ~5 × FOV</td><td>Capillary tubes</td></tr>
<tr><td>Radionuclide</td><td>^{99m}Tc</td><td>^{99m}Tc</td></tr>
<tr><td>Activity</td><td>~200 MBq</td><td>~200 MBq</td></tr>
<tr><td>Pixel size</td><td>0.7 mm</td><td>1.0 mm</td></tr>
<tr><td>Counts acquired</td><td>200 in max pixel</td><td>400 in max pixel</td></tr>
<tr><td>Analysis</td><td colspan="2">Draw several 25 mm wide profiles across lines and calculate FWHM and FWTM of each line</td></tr>
<tr><td rowspan="2">Notes</td><td colspan="2">Perform with phantom or source in both the x and y direction</td></tr>
<tr><td></td><td>Perform on the collimator surface and at distances of 5 cm and 10 cm</td></tr>
<tr><td>Frequency</td><td>Acceptance testing
Annually
If other test results are poor</td><td>Acceptance testing
Annually
After collimator damaged</td></tr>
</table>

3.11.4 Resolution (quantitative) using a point source

For full details of intrinsic measurement, see Section 3.4.4 and for system measurement, Section 3.4.6

Parameter	Intrinsic measurement	System measurement
Collimator	None	All available
Source	Collimated lead pots with 1 mm holes	Array of 1 mm dimples in Perspex sheet
Radionuclide	^{99m}Tc	^{99m}Tc
Activity	~300 MBq in each pot	1 drop of concentrated eluate in each point
Pixel size	0.7 mm	1.0 mm
Counts acquired	2000 in max pixel	
Analysis	Draw wide profiles through each point and calculate FWHM and FWTM in both x and y directions	
Notes	Repeat with sources in several different positions	Perform on the collimator surface and at distances of 5 cm and 10 cm
Frequency	Acceptance testing Annually If other test results are poor	Acceptance testing Annually After collimator damage

3.11.5 Spatial linearity

For full details, see Section 3.5

Parameter	**Intrinsic measurement**
Collimator	None
Transmission phantom	Slit phantom or orthogonal hole phantom
Source	1 mL at distance of $\sim 5 \times$ FOV
Radionuclide	^{99m}Tc
Activity	~200 MBq
Pixel size	0.7 mm
Counts acquired	1 million total
Analysis	Draw profiles and determine position of each line or point. Fit these to a straight line and determine the deviation of each line or point from the fit. Find maximum deviation
Notes	If using slit phantom, repeat in orthogonal position If using hole phantom, draw profiles in orthogonal direction
Frequency	Acceptance testing If other tests show unexpected results

3.11.6 Count rate performance

For full details, see Section 3.6.1

Parameter	Intrinsic measurement
Collimator	None
Source	1 mL at distance of ~1 m
Radionuclide	^{99m}Tc
Activity	~50 MBq (or enough to achieve saturation)
Count rate	Initially saturated (>200 000 cps) decaying to <1000 cps
Pixel size	128 × 128 matrix
Images acquired	5 minute static every hour for 60 h without moving source Final 5 min background with source removed
Analysis	Subtract background count rate from each measurement Decay correct each measurement back to time zero Determine initial expected count rate from plateau of graph Calculate expected count rate at time of each image Plot observed against expected rate and find where 20% loss occurs
Frequency	Acceptance testing After major component replacement

3.11.7 System sensitivity

For full details, see Section 3.7

Parameter	**System measurement**
Collimator	All available
Source	Thin phantom placed on collimator face
Radionuclide	^{99m}Tc
Activity	~20 MBq (accurately measured using radionuclide calibrator and corrected for decay and residual activity in syringe)
Pixel size	128 × 128 matrix
Images acquired	2 minute static image and 2 min background image
Analysis	Subtract background and calculate cps from phantom Deduce sensitivity in cps/MBq
Frequency	Acceptance testing Monthly

3.11.8 Energy resolution

For full details, see Section 3.8

Parameter	**Intrinsic measurement**
Collimator	None
Source	1 mL at distance of ~5 × FOV
Radionuclides	^{99m}Tc and ^{57}Co
Activity	~10 MBq
Count rate	<20 000 cps
Images acquired	No images. Acquire energy spectrum of each radionuclide with 10 000 counts in peak channel
Analysis	Determine energy calibration from separation of photopeaks Determine energy resolution from FWHM of ^{99m}Tc photopeak
Frequency	Acceptance testing Annually

3.11.9 Multiple window spatial registration

For full details, see Section 3.9

Parameter	Intrinsic measurement
Collimator	None
Source	Collimated thick lead pot with 2 mm hole
Radionuclide	^{67}Ga
Activity	~50 MBq
Images	Separate image in each of three windows, 93, 185 and 300 keV
Pixel size	0.7 mm
Counts acquired	1000 counts in maximum pixel
Analysis	Calculate centroid of source position in each energy window Determine maximum difference between any of the windows
Notes	Repeat with source at several locations within FOV
Frequency	Acceptance testing Annually

4 Quality control for whole-body imaging

Glen Gardner

4.1 Introduction

Whole-body imaging involves the movement of either the patient bed or the camera head and, as such, introduces new sources of error and a requirement for further quality control (QC) measurements. For the production of good clinical images, the movement must be both uniform and smooth, requiring good motor control of the bed or gantry. Several publications have discussed the topic of QC of whole-body imaging systems, most notably BSI (1998), and Blokland et al. (1997), and more recently, the Medical Devices Agency Gamma Camera Assessment Team (MDA, 2001) and NEMA (2012).

The tests discussed here compliment those undertaken for planar imaging and can therefore be performed less frequently. They should none the less be undertaken often enough to ensure optimal whole-body imaging performance.

The tests described investigate the whole-body performance during continuous motion acquisition. They could be adapted easily to test 'step-shoot-stitch' mode with careful placing of the phantoms to span the join or stitch area. In continuous motion, it is important to test the integrity of the ramp area at the beginning and end of the scan. This is necessary to produce even counting statistics across the whole image.

4.2 Comparison with planar performance

The first two tests described here should be performed at acceptance testing. They establish a baseline comparison of whole-body versus static planar values for non-uniformity and spatial resolution.

4.2.1 Whole-body system non-uniformity

This test uses the scanning camera to image a stationary ^{57}Co flood source (or other fillable flood, see Section 3.3.2) positioned on the camera bed. The speed of camera movement and activity of the flood source will inevitably limit the total number of counts in the final image.

In order to optimise the counts acquired, the most sensitive low-energy parallel hole collimator is used along with the normal default scanning speed (usually 10 cm min^{-1}). The source is placed on the bed in the centre of the scanned area (i.e. avoiding the ramp areas), orientated such that the long axis is parallel to the long axis of the camera head, assuming that the source is small enough to pass through the gantry ring.

The resulting image is analysed to gain the total counts within an area corresponding to one field of view; this area is equivalent to that of a static image of the same flood source. A second scan is undertaken by positioning the camera over the source (taking care not to alter the head to source distance) and obtaining a static planar image containing the same number of counts as within the corresponding region on the whole-body image. Images are acquired into a 256 × 256 matrix for the static image and a 1024 × 256 matrix for the scanned image; the central section of the scanned image containing the flood data is extracted into a 256 × 256 matrix.

Analysis similar to that described in Section 3.3.4 can then be performed on the flood images. Both flood images are rebinned into 64 × 64 matrices to increase the pixel statistics. Regions of interest (ROIs) are created from the static image for useful (UFOV) and central field of views (CFOV), and these ROIs subsequently used on the image from the whole-body scan. Measurements of integral and differential uniformity as well as coefficient of variation (CoV) are made following the procedures detailed for planar system uniformity tests. The test should be repeated on all camera heads and the results analysed. Analysis of the static and scanned images is both quantitative and qualitative to determine any significant difference in performance, notwithstanding any limitations because of low count statistics. Depending on the activity of the source and the sensitivity of the detectors, there will be statistical limitations because of the low counts in the acquired image. However, the quantitative measurements should be within ±1% of each other with the possibility that the scanned image will show better uniformity owing to the smoothing effect of the bed motion.

4.2.2 Whole-body system spatial resolution

An emission line source phantom (Figure 4.1) can be used to assess the spatial resolution of the total detector system while acquiring a whole-body image; this can then be compared to a static image with the same phantom/camera set-up. Data obtained are analysed to determine the full width at half maximum height (FWHM) of the line spread function (LSF) of a series of profiles taken from each emission line source.

As the aim of this test is solely to compare the camera resolution during movement against that of a static scan, a wide variety of line source phantoms could be used – provided that a sufficiently high count rate can be achieved and that the phantom will physically fit through the gantry during a whole-body acquisition. This test differs from the planar resolution tests described in Section 3.4 and aims to give an indication of the resolution that might be expected from a whole-body patient scan.

The phantom described here is based on that used for the Medical Devices Agency (MDA) assessment of planar system resolution (Bolster et al., 1996), with the

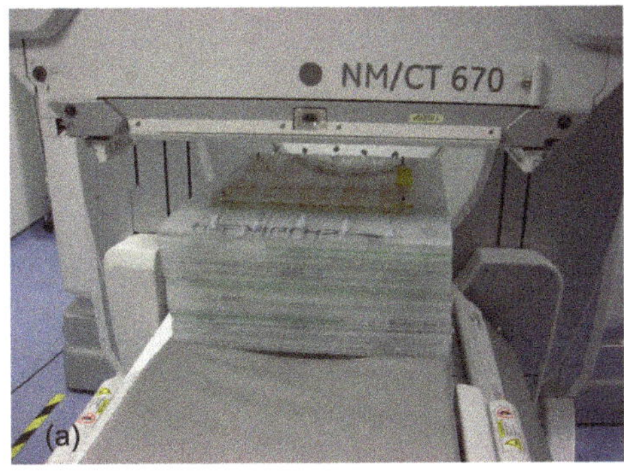

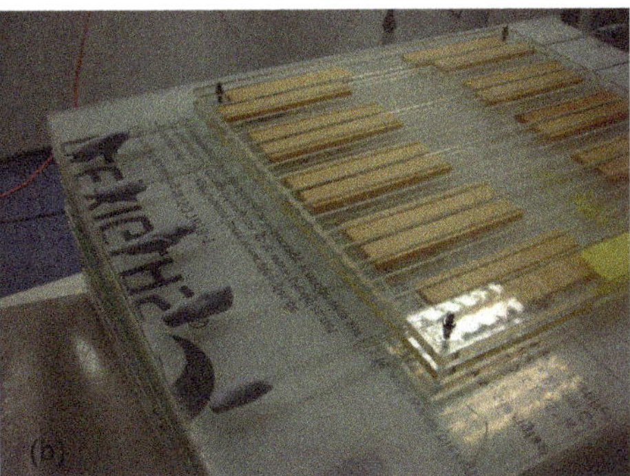

Figure 4.1 Photographs of the whole-body resolution phantom (note: images shown with only 10 mm Perspex on top as scatter). (a) The glass capillary tubes are shown positioned within parallel slots between two pieces of Perspex. Blu-Tack® is used to seal the ends. (b) The whole-body resolution phantom within 200 mm of scatter

dimensions reduced to enable the phantom to pass through the centre of modern slip ring cameras (Figure 4.2). Simpler designs may be adequate for the requirements of the user. The five line sources comprise glass capillary tubing with a 0.5 mm bore diameter. These are positioned within parallel slots cut into a Perspex base sheet with a centre-to-centre slot separation of 60 mm. The capillary tubes are filled with ^{99m}Tc solution and sealed at both ends. The total activity present within the field of view is made to be such that the count rate is approximately 2×10^4 cps.

In order to mimic the clinical situation, measurements are performed with 150 mm scattering media interposed between the line sources and the collimator surface. Additional sheets are used to provide backscatter, bringing the total thickness of the assembly to 200 mm. The phantom is placed on the bed and aligned with the x axis (perpendicular to the bed) of the camera, using the persistence scope as a guide to alignment. The camera head is then brought as close as possible to the phantom surface while still enabling the whole-body scan to proceed. The phantom is then scanned at the speed recommended by the manufacturer for clinical images, usually around 10 cm min^{-1}. First, the phantom is placed at the start of the scanning area to test the electronic ramping accuracy. Second, the phantom is placed at the centre of the scanning area and a repeat measurement made. Finally, a static acquisition is obtained at the same camera–phantom separation, with the number of counts acquired set equal to that obtained during the whole-body scan in one field of view. These procedures are then repeated with the phantom orientated in the y direction.

Data are collected into the maximum possible consistent pixel matrix, usually 1024×256 for the whole-body image, and 256×256 for the static image. Data are

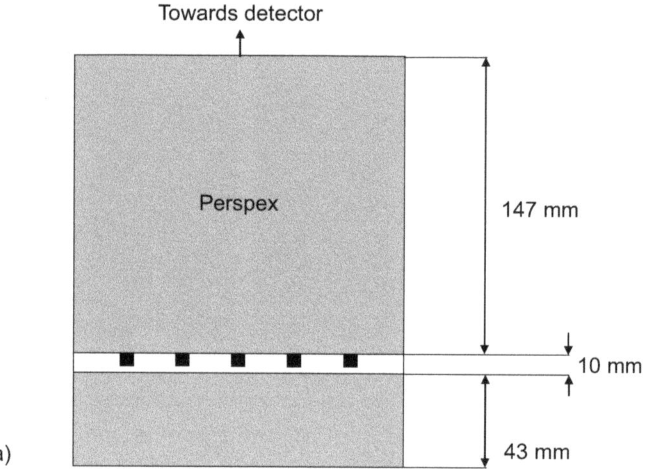

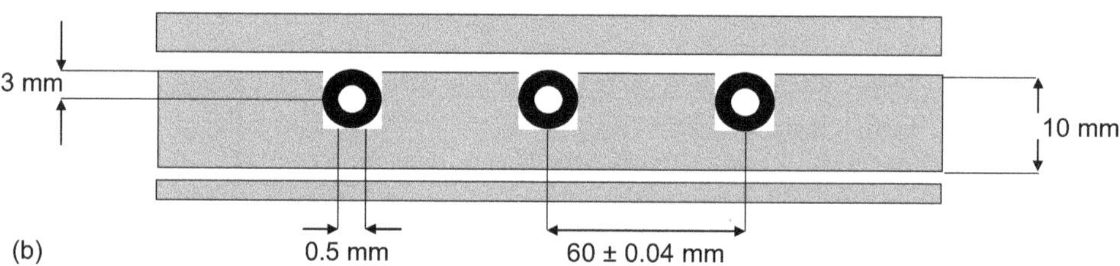

Figure 4.2 Detail through whole-body resolution phantom showing construction of slit channels. (a) View of the phantom plus Perspex scatter; 147 mm of Perspex sits between the phantom and the imaging detector; the 43 mm of Perspex is used as backscatter. (b) A closer view of the phantom, showing the glass capillary tubes situated in the Perspex grooves. This phantom design is illustrative; simpler designs may also be used

subsequently reformatted into 16 profiles perpendicular to the axis of the lines and the mean values for FWHM and full width at tenth maximum (FWTM) height are generated in millimetres, using the known line source separation to convert from pixels to millimetres.

A comparison is then made between the results from the scanned ramp position, the central scanned position, and the static image for both the x and y orientations (Table 4.1). There should be no significant differences between the whole-body and static image resolution.

Table 4.1 Typical example of resulting measurements (mm) for a low-energy general purpose (LEGP) collimator scanning at 10 cm min^{-1} with a count rate of 6 kcps, using 150 mm Perspex for scatter

Orientation		**Y**		**X**	
Scan area	**Ramp**	**Middle**	**Static**	**Middle**	**Static**
FWHM mean	16.75	16.79	16.23	16.26	16.28
FWHM SD	0.80	0.58	0.68	0.66	0.82
FWHM CoV	4.81	3.47	4.20	4.04	5.03
FWTM mean	30.52	30.60	25.59	29.64	29.68
FWTM SD	1.47	1.06	1.24	1.20	1.49
FWTM CoV	4.80	3.47	4.20	4.04	5.03

4.3 Whole-body count rate variation

This test can be undertaken at acceptance testing but is straightforward and could therefore be incorporated into a routine quality assurance (QA) programme. The frequency of testing will clearly depend on the number of whole-body procedures being undertaken, but in a routine department using whole-body imaging regularly, it is recommended that the test be undertaken either monthly or quarterly.

A point source positioned on the scanning camera head should result in an image containing a straight line of constant count value; any variation in this count may indicate a fault in either the mechanical speed of the detector/bed movement or a fault in the positioning or timing electronics.

A ^{99m}Tc point source containing sufficient activity to produce a count rate of around 20 000 cps is attached to the camera at the centre of the collimator face. A drop of high specific activity at the end of a needle cap could be used for the point source. Whole-body scans are then performed over the maximum length at clinically relevant speeds, usually ranging from 10 to 20 cm min^{-1}, acquiring the images into a 256×1024 matrix and using the highest sensitivity parallel hole collimators available. High sensitivity collimators are used to allow a smaller volume whilst retaining the desired 20 000 cps. Depending on the particular camera system, however, care must be taken to avoid pixel saturation, and a count rate of significantly less than 20 000 cps may be required.

The acquired image is then cropped to exclude the ‘hot spots’ at the ends of the line, which are an artefact created by the ramp process. The image data are rebinned into a 64 × 256 matrix and a profile created along the line, with the counts corrected for decay during the scan. Calculations are performed in order to obtain the mean, standard deviation (SD), and CoV of the pixel values along the profile.

The measurements from the profile are examined both qualitatively for visual abnormalities and quantitatively for statistical irregularities. Figure 4.3 illustrates a type of fault that can be detected with this test. It is important that the source used is sufficiently small to ensure that any irregularities are not smoothed out by the scanning motion. Quantitatively, the measured CoV should be around 1% for a 10 cm min^{-1} scanning speed with a count rate of 20 000 cps.

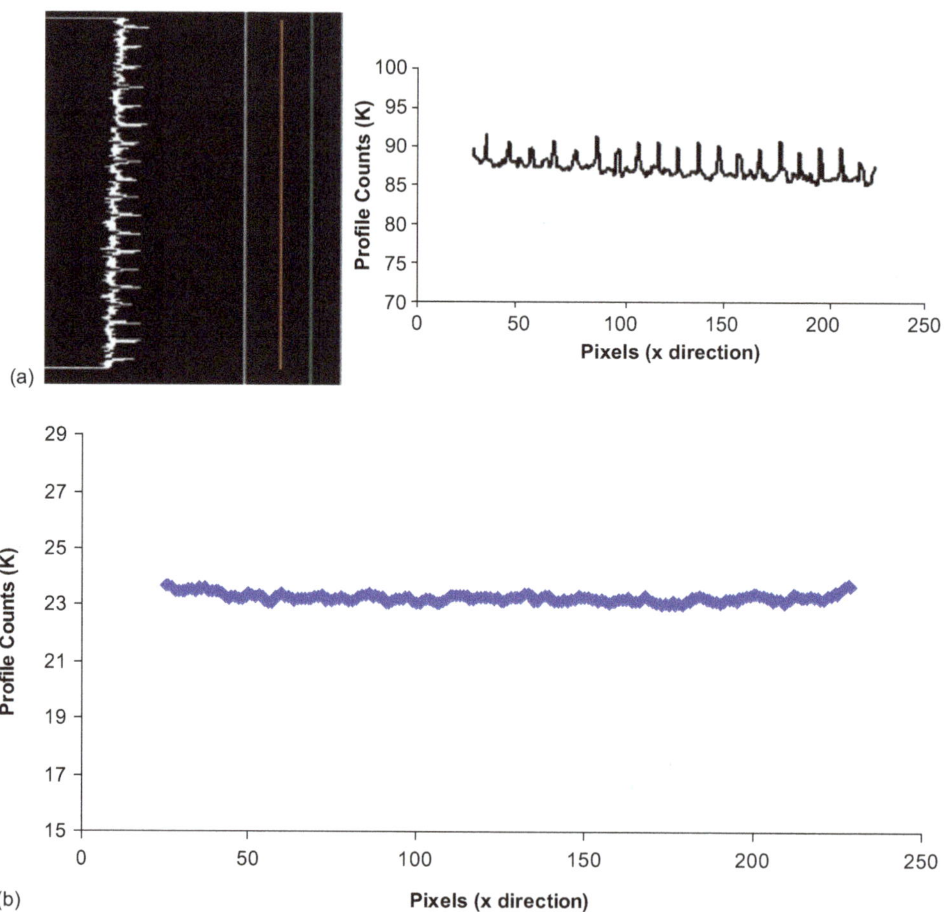

Figure 4.3 (a) Whilst the source is stationary and independent from the bed motion, this artefact points to either a fault with the digital encoder or the bed motor mechanics. (b) This profile is representative of acceptable performance

4.4 Whole-body exposure time correction

This test can form part of a routine QA programme, since it is straightforward to undertake. All areas within a whole-body image should have equal exposure times, hence the requirement for electronic ramping at the start and end of the scan. A simple and effective way to ensure the accuracy of this exposure time correction is to attach a solid ^{57}Co flood or fillable ^{99m}Tc flood to the camera head and perform a whole-body scan for the maximum possible length. The flood source should be large enough to ensure that the entire field of view is covered. The highest sensitivity parallel hole collimator available should be used to maximise the acquired counts.

The resulting image is rebinned from 256×1024 into 64×256, and columns and rows are summed to give profiles across the x and y directions. The resulting profiles should be flat from edge to edge and any abnormalities noted. Quantitative measurements of the mean and SD of counts for the x and y directions, respectively, can be calculated, avoiding any ramp or edge areas.

Information on the effect of the scanning bed (Figure 4.4a) is also obtained with the reduction in counts through the bed expressed as a percentage of the non-attenuated counts. Some cameras have rectangular fields of view with cut-off corners, as in the case shown in Figure 4.4b. This results in areas of reduced counts at the edges of the image. In this case, only the non-reduced areas should be analysed quantitatively.

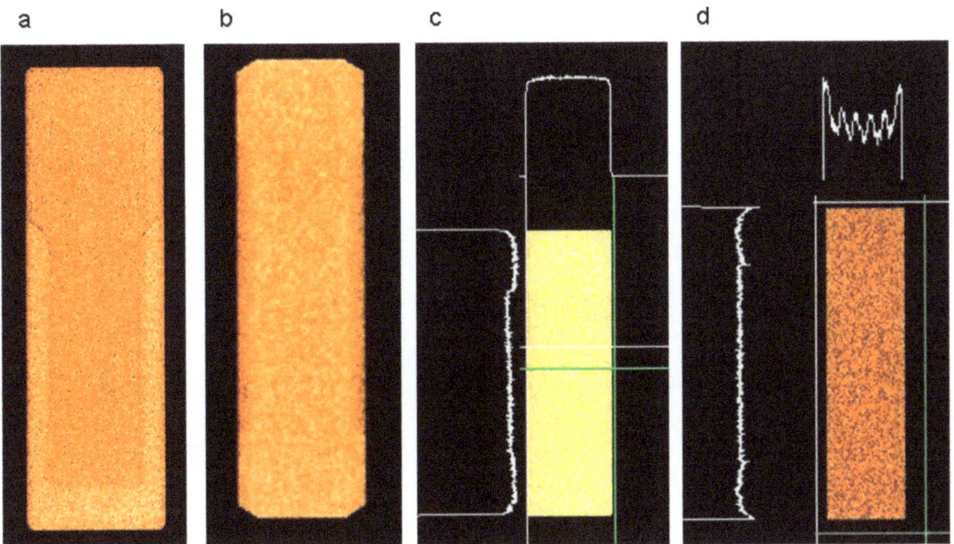

Figure 4.4 (a) Whole-body scan showing variation caused by the patient couch. (b) Edge effects due to cut-off corners on the detector. (c, d) Examples of scan profiles

Some manufacturers do perform a correction on the areas with reduced count rates, and these cameras can be treated as normal full field cameras. Figure 4.4c illustrates the variation in exposure time that can be found, with a sharp drop in counts after the initial ramp phase, indicating that the bed position decoder is non-linear.

With most detectors having photomultiplier tubes (PMTs) arranged in a hexagonal pattern, the effect of any non-uniformity is smoothed out as the detectors move down the bed. In Figure 4.4d however, the PMTs are square and the resultant whole-body scan magnifies any small non-uniformities which can be seen as vertical stripes. Therefore, for this particular detector design, it is of greater importance to review both planar and whole-body uniformity performance.

4.5 Summary of measurement protocols

For easy reference, this section contains a summary of the acquisition parameters recommended for each of the measurements described in this chapter. Full details can be found in the relevant sections.

4.5.1 Whole-body system non-uniformity

For full details, see Section 4.2.1

Parameter	System measurement
Collimator	LEGP
Source	Uniform flood source
Radionuclide	^{57}Co
Pixel size	256 × 1024 matrix (whole-body) 256 × 256 static
Images acquired	Whole-body scan with the source in the centre of the scan Static scan of the source, with the same detector distance as the whole-body. Repeat for each detector being tested
Scan speed	Default (10 cm min^{-1} or 10 min m^{-1})
Analysis	Extract the flood image from the whole-body scan (256 × 256 matrix) Convert both flood images to 64 × 64 matrices Analyse in same way as planar uniformity Compare
Frequency	Acceptance testing After major component replacement

4.5.2 Whole-body resolution

For full details, see Section 4.2.2

Parameter	**System measurement**
Collimator	LEGP
Source	Capillary tubes
Radionuclide	^{99m}Tc
Count rate	~20 000 cps
Pixel size	256 × 1024 matrix
Activity	~500 MBq in 1 mL
Images acquired	150 mm of scatter between detector and capillary tubes Whole-body image with phantom in ramp region. Whole-body with phantom in centre of scan. Static image without altering detector–phantom distance
Scan speed	Default (10 cm min^{-1} or 10 min m^{-1})
Analysis	Analyse in same way as planar resolution Compare static and whole-body results
Frequency	Acceptance testing After major component replacement

4.5.3 Whole-body count rate variation

For full details, see Section 4.3

Parameter	System measurement
Collimator	LEGP
Source	Point source (drop of activity in needle cap)
Radionuclide	^{99m}Tc
Count rate	~20 000 cps
Pixel size	256 × 1024 matrix
Activity	~100 MBq (@2000 MBq mL^{-1})
Images acquired	Whole-body scan with the point source attached to the centre of the collimator face
Scan speed	Default (10 cm min^{-1} or 10 min m^{-1})
Analysis	Convert image to a 64 × 256 matrix Obtain a profile avoiding the ramp regions Determine the mean, standard deviation, and CoV from the pixels in the profile
Frequency	Acceptance testing Quarterly

4.5.4 Whole-body exposure time correction

For full details, see Section 4.4

Parameter	System measurement
Collimator	LEGP
Source	Uniform flood source
Radionuclide	^{57}Co
Pixel size	256 × 1024 matrix
Images acquired	Whole-body scan with the uniform flood source placed on top of the detector. Repeat for each detector being tested
Scan speed	Default (10 cm min^{-1} or 10 min m^{-1})
Analysis	Convert image to a 64 × 256 matrix Obtain profiles in the x and y directions avoiding ramp or edge areas Determine the mean, standard deviation, and CoV from the pixels in the profile
Frequency	Acceptance testing Monthly

5 Assessment of performance for single photon emission computed tomography

Original by Peter Jarritt and Wendy Waddington, 2003
Revised by Richard Lawson, 2013

5.1 Introduction

The extension from planar imaging to single photon emission computed tomography (SPECT) imaging imposes more stringent requirements on the acceptance testing, calibration, quality control (QC), and clinical use of a gamma camera. SPECT can amplify any small camera non-uniformities that lie close to the axis of rotation, and errors in the centre of rotation can blur the reconstructed image. The patient must remain still during the acquisition, and with multi-detector systems, the detectors must be properly aligned. Therefore, attention to detail is essential with SPECT QC. This chapter will describe the SPECT tests that should be carried out as part of acceptance testing of a new SPECT camera and again at suitable intervals during its lifetime. The tests discussed should be carried out in addition to the planar tests already described in Chapter 3. It is important that these tests are performed as part of a regular quality assurance programme for any gamma camera that is to be used for SPECT imaging.

Additional quality control measures required by hybrid SPECT-CT systems are described in Chapter 6. The use of radionuclide sources for transmission imaging in order to obtain an attenuation map is no longer included in this report as they have now been superseded by SPECT-CT systems. A brief discussion of the requirements for transmission imaging systems using radionuclide sources can be found in Chapter 5 of IPEM Report 86 (IPEM, 2003a).

5.2 Principles of the tomographic process

In order to explain the rationale behind a SPECT QC programme, it is necessary to understand the factors which can influence the accuracy of the tomographic process. SPECT acquisition is discussed in detail in a recent IPEM publication (Lawson, 2013) and SPECT reconstruction is described in IPEM Report 100 (Lawson, 2011). Therefore, this section will only review the main factors which can affect the quality of SPECT images. Subsequent sections of this chapter will describe how various SPECT QC tests are designed to verify that these factors do not adversely affect the quality of the reconstructed images.

5.2.1 Uniformity of response

The tomographic process comprises the acquisition of a series of projection images collected at a range of angles around the patient or object, which are then reconstructed into cross-sectional images of the patient. The method relies on obtaining a consistent and uniform response from the gamma camera detectors, regardless of where the gamma rays are detected. This requires not only a uniform response across the detector face, but also that this does not change with the angle of rotation and that, for multiple detector systems, the detectors are well matched in their response. In this context, uniformity implies more than just uniform detector sensitivity; it also requires a uniformity of energy resolution, spatial resolution, and pixel size. Variations in energy resolution can change the fraction of events falling within the energy window, which can lead to non-uniformity and also cause degradation of clinical studies due to the incorporation of differing proportions of scattered photons in different regions of the detector. Spatial non-linearity can also be a major source of image non-uniformity, as only a small deviation in event position towards the centre of a photomultiplier tube (PMT) can produce a large increase in local count density (Lawson, 2013). Variations in spatial resolution lead to loss of reconstructed resolution and also loss of contrast due to an increased partial volume effect. In SPECT, the partial volume effect occurs when a 'hot' object is smaller than the camera resolution so that it becomes blurred out to appear larger than it really is. Since the total counts due to the object are preserved, a broader image must imply that it does not look as 'hot' as it should – in other words, the contrast compared with surrounding background is reduced. Variations in pixel size will also lead to non-uniformity in the count density in projections. If attenuation correction is applied, then errors in pixel size will also lead to an incorrect allowance for the amount of attenuation expected in each pixel.

Most of these effects can be masked using sensitivity corrections from stored correction maps. These correction maps are acquired using a uniform source, and so sensitivity correction simply scales counts in each pixel up or down according to whether the uniform source appears 'cold' or 'hot' relative to the average. Therefore, sensitivity correction can always make the image of a uniform source appear perfect, no matter how non-uniform the uncorrected image appears. However, sensitivity correction does not address the fact that events may be misplaced due to camera spatial non-linearities and so simply scaling counts up or down is not sufficient if the source is non-uniform – like a patient. Therefore, corrections which address the fundamental properties of energy and spatial response are considered more important than sensitivity corrections, and cameras used for SPECT should have energy correction and linearity corrections applied in addition to sensitivity corrections. Only gamma cameras with good uniformity should be used for SPECT and therefore regular uniformity checks are an essential part of SPECT QC.

5.2.2 Alignment of projections

Tomographic reconstruction of the projection data into transaxial images has to accurately mirror the acquisition process by re-projecting the data from the equivalent angles and positions that were used in the acquisition process. These depend on the mechanical rotation of the gamma camera detector (or detectors) about their axis. The mechanical rotation defines a line in space called the axis of rotation. In general, if a point source is imaged by the detectors, then as the detectors rotate, the position of the source in the projection image will appear to wobble from side to side in a sinusoidal motion. However, if the source is placed on the axis of rotation, then its image will remain in exactly the same position as the camera rotates. This position is called the centre of rotation (COR) and for accurate reconstruction, the reconstruction software needs to know exactly where the COR lies within the image in the X direction. By default, the COR is usually assumed to be at the centre of the image, i.e. between pixels 32 and 33 for a 64×64 matrix or between pixels 64 and 65 for a 128×128 matrix. If the COR is not exactly in the middle of the image in the X-direction, then a COR offset must be defined to specify the actual location of the COR relative to the centre of the image (see Figure 5.1). The SPECT reconstruction software usually assumes that the COR is at the centre of the image and so if there is a non-zero COR offset defined, then it must shift each acquired projection by the correct amount before reconstruction starts. Some gamma cameras will therefore store SPECT data with an appropriate COR offset stored in the header information, whereas others will correct the raw event position data in real time and therefore set the COR offset to zero. Therefore, verification that the stored COR offset is correct is an essential part of SPECT QC.

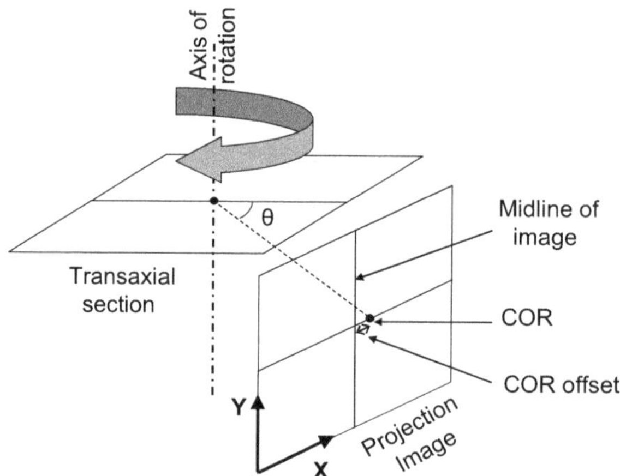

Figure 5.1 COR offset is defined as the distance between the midline of the image in the X direction and the point where a source on the axis of rotation would be imaged

The reconstruction software also assumes that all of the projection images are equally spaced around the acquisition arc, but in practice, small mechanical errors will exist. In 'step-and-shoot' mode, if the detector fails to stop at exactly the correct angle, this will lead to degradation in the reconstructed images. Moreover, all mechanical systems are subject to a degree of hysteresis, which means that the actual detector position will depend on whether it is rotating clockwise or anticlockwise. In continuous rotation mode, the rotational speed throughout the study should be consistent – which may not be easy to achieve if the gantry is unbalanced with the weight of detectors all on one side.

5.2.3 Stability of response with rotation

The requirements for a consistent and uniform response and the correct alignment of projections must pertain at all rotational positions of the detector. The detection process of most gamma cameras utilises many photomultiplier tubes (PMTs) which are easily affected by stray magnetic fields. The PMT works by accelerating electrons in a vacuum from the photocathode to the first dynode and then between successive dynodes. Any magnetic field will deflect the electron trajectory and hence alter the gain of the PMT. Even the weak magnetic field of the Earth is sufficient to affect a PMT gain and so, as a gamma camera detector rotates around the patient, the changing orientation relative to the Earth's magnetic field can affect the uniformity of the image. Manufacturers employ various techniques to minimise this effect, for example, mu-metal shields around the PMTs and circuits for automatic gain stabilisation of each PMT.

Detector orientation can also have an effect on air currents within the detector housing and this can cause temperature variations which could affect the operation of detector electronics. Even gravitational effects could affect the collimator position if it is not mounted securely. Therefore, for SPECT, it is important to know that the detector performance does not change with rotation angle.

5.3 Planar uniformity

Good planar uniformity is essential for accurate SPECT studies and so the uniformity of each detector should be monitored regularly using one of the methods described in Section 3.3. Since collimator damage can affect the reconstructed results, it is preferable to assess system uniformity rather than intrinsic uniformity and each collimator that will be used for SPECT should be tested regularly. Planar uniformity should be performed as one of the acceptance tests for any new SPECT camera. Thereafter, daily QC should include at least a visual inspection of a 5 million count flood image. At least once a week, a full quantitative assessment of a high count uniformity image should be made as described in Section 3.3.5. Results should be plotted on a control graph (Section 9.6.2) and if they deviate beyond acceptable limits, then consideration must be given as to whether the system may continue to be used for SPECT. It is possible that the system may still be acceptable for planar imaging even if it cannot be used for SPECT. Action to be taken if the uniformity is deemed unacceptable will depend on the camera design, and the manufacturer's recommendations should be followed. Remedial action might include retuning of the detector, acquisition of a new sensitivity correction map (which can often be done by the user) or new energy and linearity correction maps (which will probably require a visit from the service engineer).

5.4 Global sensitivity

It is valuable to combine a measure of global sensitivity with the daily assessment of planar uniformity. This is most easily achieved if the measurement is made with a long-lived source such as ^{57}Co. Provided that the imaging conditions are constant, the total count rate in the whole field of view should only fall from day to day in line with the known half-life of ^{57}Co (272 days). This test can highlight drifts in detector peaking or detector ageing or PMT decoupling.

Action thresholds should be defined for changes of more than $\pm 2\%$ from the expected value.

5.5 Centre of rotation

A centre of rotation test is performed to check that the current value of the stored COR offset is still valid. This is not necessarily the same as the manufacturer's procedure for acquiring a new COR offset which will overwrite the current COR offset with a new value. It is advisable to monitor the COR position on a regular basis, but only replace the COR offset with a new value if the current one is found to be out of specification. In that way, one can get an appreciation of how stable the COR is and hence adjust the measurement intervals accordingly. For a modern camera, it should be adequate to check the COR for each collimator on a monthly basis and new COR values will probably only need to be saved after a service or other mechanical or electronic intervention. If a new COR offset is saved every month as a routine, then there may be no indication of whether this was really necessary or not. If the manufacturer's procedure allows the user to compare the new COR offset with the current one before deciding whether to save the new value or keep the current one, then the manufacturer's procedure may be used as a checking test without saving the new value. However, if the manufacturer's procedure does not allow comparison of the new value with the current one, then it is better to use a separate check procedure as described in this section. If the manufacturer's software is to be used, then the user should be fully aware of what it does and what the results mean.

Some manufacturers include a centre of rotation check as part of a more comprehensive test using multiple point sources and special analysis software.

However, the tests described here can be carried out using only simple sources and analysis software that can be written by the user. The centre of rotation test is designed to validate the electrical and mechanical alignment of each detector with the axis of rotation. This can be affected by mechanical movement of the detector or the gantry, rotation of the detector within the head, electronic shift of the X or Y axes in the detector, or damage to holes in the collimator. The methods described here are based on those recommended by the International Atomic Energy Agency (IAEA, 2009).

5.5.1 Head levelling check

For SPECT, it is important that the detector heads are level – or at least parallel to the axis of rotation which is assumed to be horizontal. There are two ways to check this: one method can be incorporated into the COR check described in Section 5.5.2 and the other is a mechanical check described here. This mechanical check is suitable for an acceptance test, but the COR check is more suitable for routine use.

The gantry should be set up as if to perform a standard SPECT acquisition and then rotated so that one detector is positioned at the bottom of its orbit with the collimator facing up. A spirit level should then be placed on the collimator surface to verify that the detector is horizontal (in the direction parallel to the axis of rotation). Then the gantry should be rotated until the same detector is at the top of its orbit and the spirit level tested against the collimator again. If the collimator remains horizontal in both positions, then all is well. If the collimator shows the same tilt and in the same direction in both positions, then this indicates that the axis of rotation is tilted but the detector is parallel to the axis, which is acceptable. If the tilt is different in each position, then this indicates that the detector is not parallel to the axis of rotation. In this case, if there is no user adjustment of the tilt angle, the manufacturer should be contacted for advice. For multiple headed systems, the test should be performed for each detector separately. Each detector should be parallel to the axis of rotation within 0.5°, which is just about the limit of detection of an ordinary spirit level.

5.5.2 Centre of rotation check

The centre of rotation check requires a small volume source – ideally a point. NEMA recommends using a drop of less than 2 mm size (NEMA, 2012) but it is difficult to get sufficient activity in this volume. So, since the analysis will determine the centroid of the activity distribution, it is more practical to use a larger volume as long as this is reasonably symmetric. A volume of 0.1 mL in the barrel of a 1 mL syringe (not including the needle) gives a cylinder 5 mm long and about 5 mm diameter which is adequate. An activity of 20–50 MBq ^{99m}Tc in less than 0.1 mL should be drawn up into a 1 mL syringe and then the needle removed and replaced with a blind hub. The syringe should then be taped to the end of a plastic or wooden rod and positioned so that it projects over the end of the patient couch as illustrated in Figure 5.2. The source should be positioned towards the centre of the field of view in the axial (Y) direction but offset from the axis by about 10 cm towards one of the detectors. If the source is placed close to the axis of rotation, then the COR can still be measured, but any detector tilt will not be detected, so it is preferable to place the source off-axis.

A normal tomographic acquisition should be performed for each head independently, and for multiple detector systems, the source must not be moved between measurements on each head. An acquisition consisting of 30 projections over 360° is adequate. Note that each detector must acquire a full 360° arc, so a typical patient protocol with a dual head acquisition where each head contributes 180° is no good for this test. A 128 × 128 matrix and an acquisition of about 10 s per view will give adequate counts.

If the collimator holes are not exactly perpendicular to the collimator face, then the measured centre of rotation can vary with the distance of the collimator from the axis and so the radius of rotation should be set at a typical small patient position and repeated for a large patient position. If the crystal is not mounted exactly square in the detector, then the electronic Y axis may not be exactly parallel to the

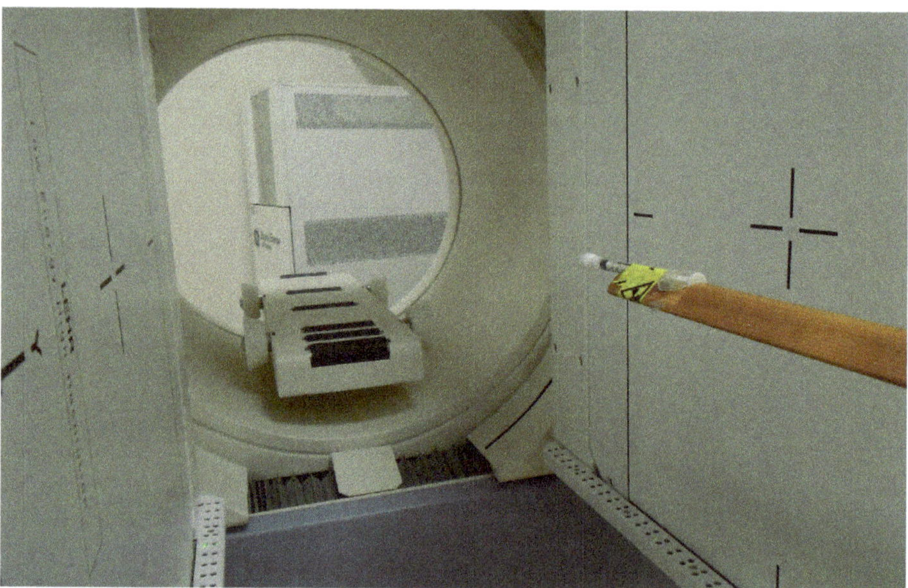

Figure 5.2 A source suitable for checking the centre of rotation can be made from activity in the end of a 1 mL syringe attached to a wooden pole projecting from the end of the patient couch. Note that the source is deliberately positioned slightly off axis towards one detector

mechanical axis of rotation. In this case, the centre of rotation can vary with the axial position and so ideally, the measurements should be repeated with the source at several positions along the axis (Y direction of the image). Of course, the source must remain within the field of view of the detector at all rotation angles. If cardiac SPECT is performed, then the test should be performed in perpendicular mode (with detectors 90° or 76° apart) as well as parallel mode (with detectors 180° apart). To fully check the system, the test should be repeated separately for both clockwise and anticlockwise rotations of the gantry and for circular and non-circular orbits if these are used.

Analysis of the data acquired in this way requires the determination of the location of the centre of the source in each projection image. This can be calculated from the centroid of image counts in both X and Y directions. If $C(i,j,\theta)$ represents the counts in the pixel in the ith row and jth column of the projection image acquired at a rotation angle θ, then the coordinates of the source centre, $(X_S(\theta),\ Y_S(\theta))$, can be calculated using

$$
\begin{aligned}
X_S(\theta) &= \frac{\sum_i \sum_j (i-0.5)p_x C(i,j,\theta)}{\sum_i \sum_j C(i,j,\theta)} \\
Y_S(\theta) &= \frac{\sum_i \sum_j (j-0.5)p_y C(i,j,\theta)}{\sum_i \sum_j C(i,j,\theta)}
\end{aligned}
\qquad (5.1)
$$

Here p_x and p_y are the size of a pixel (in mm) in the X and Y direction, respectively. The coordinates i and j are assumed to run from 1 to 128 so that $(i\text{–}0.5)\ p_x$ is the X position of the middle of the ith pixel. If i and j run from 0 to 127, then this should be replaced in Equation 5.1 with $(i+0.5)\ p_x$. The summations over i and j do not need to include all rows and columns in the image but should be limited to just a range of about 25 rows centred on the source. This should be sufficient to include all significant counts from the source.

The results of this calculation should be plotted as graphs of $X_S(\theta)$ and $Y_S(\theta)$ against rotation angle θ. Ideally, $X_S(\theta)$ should vary in a sinusoidal manner with a mean value equal to the centre of rotation. Therefore, the $X_S(\theta)$ data should be fitted to a sinusoidal function

$$X_{Fit}(\theta) = X_{COR} + A\ Sin(\theta + \phi) \qquad (5.2)$$

If the data were acquired with equal increments over a full 360° arc, then the mean fit value, X_{COR}, will just be equal to the average of all $X_S(\theta)$ values, but if the acquisition arc is limited (as happens with some dedicated cardiac cameras), a formal fitting routine will be needed to determine X_{COR}. The fitted value of X_{COR} (in mm) should be compared with the midpoint of the image matrix plus any currently defined COR offset. The two should agree to within ± 1 mm. The amplitude of the sinusoid, A, merely reflects the distance that the source was placed away from the axis of rotation. The deviation of the measured source position from the sinusoidal fit at each angle, $X_S(\theta)$–$X_{Fit}(\theta)$, should also be plotted as this gives an indication of how much the COR offset varies with angle. If the system uses only a single COR offset for all angles, then the maximum deviation from the fit should not exceed ± 1 mm.

Ideally, $Y_S(\theta)$ should be independent of rotation angle so the $Y_S(\theta)$ data should be fitted to a constant.

$$Y_{Fit}(\theta) = Y_{Pos} \tag{5.3}$$

The measured value of the mean source position, Y_{Pos}, is not significant, but for a multiple headed system, it should be the same for all detectors within ± 2 mm.

If $Y_S(\text{Max})$ and $Y_S(\text{Min})$ are the maximum and minimum values of the source position taken over all angles, then the detector tilt, α, can be determined using (Lawson, 2013)

$$\text{Sin}(\alpha) = \frac{(Y_s(\text{Max}) - Y_s(\text{Min}))}{2A} \tag{5.4}$$

where A is the offset of the source in the X direction determined from Equation 5.2. The tilt angle should be less than 0.5°. Note that, if the source is placed close to the axis of rotation, then A is zero and then α cannot be determined. Therefore, despite the fact that some manufacturers suggest putting the source close to the axis of rotation, it is better to place it off-axis.

Protocol for centre of rotation check

a) Fit the collimators used for SPECT and put the detectors into either perpendicular or parallel mode as they would be for clinical studies.

b) Prepare a source of 20–50 MBq ^{99m}Tc in a volume of less than 0.1 mL in the barrel of a 1 mL syringe.

c) Suspend the source between the detectors but beyond the end of the patient couch and displaced about 10 cm away from the axis of rotation.

d) Acquire a SPECT study for each head separately with 30 views over a full 360° arc with 10 s per view on a 128×128 matrix.

e) Analyse the images for each head separately. Determine the X and Y position of the centre of the source at each angle using Equation 5.1.

f) Plot the X position of the source as a function of rotation angle. Fit this to a sinusoid and determine the mean position of the fit. Verify that this agrees with the recorded centre of rotation offset for each detector within 1 mm.

g) Determine the maximum deviation of the X position from its sinusoidal fit which should be less than 1 mm.

h) Plot the Y position of the source as a function of rotation angle and find the average value. Verify that this value is the same for both detectors within 2 mm.

i) Use the difference between the maximum and minimum Y position values to calculate the head tilt using Equation 5.4. This should be less than 0.5°.

5.5.3 Frequency of centre of rotation check

Centre of rotation must be checked as part of acceptance testing for a new camera before any other SPECT tests are performed. Thereafter, it should be checked once a month for each set of collimators that are used for SPECT. An additional centre of rotation check should be performed after any service or repair.

5.6 Collimator assessment

The construction of collimators (particularly those comprising preformed foil sheets) is prone to error, and they may also suffer damage during transportation. This can lead to a misalignment of the collimator holes and effectively leads to an error in the mechanical alignment of the detector system. For this reason, collimators made with cast construction are highly recommended for use with SPECT systems because they can be made more accurately and are more robust – although they can still become damaged during use if not treated carefully.

In addition, inaccuracies may occur in the mechanical mounting of collimators, and/or in the mounting of the collimator insert into its frame. Parallel-hole collimators can be assessed for these problems using the centre of rotation check already described in Section 5.5. Collimator errors will be revealed by changes in the COR offset with position and/or deviation from the sinusoidal fit. The failure to obtain the same COR value for all Y positions is almost certainly related to the integrity of collimator construction on modern camera systems.

The COR measurement should be performed with an offset point source at a series of positions along the detector face and the axis of rotation. The COR offset value should not change by more than 1–1.5 mm along the axis of rotation. All collimators that might be used for SPECT should be tested as COR offsets may be different for each collimator. Each set of collimators should be tested in the acquisition modes that are used clinically, i.e. parallel or perpendicular (cardiac) mode.

5.6.1 Collimator hole angulation

Another simple test of correct collimator hole alignment is the collimator hole angulation test. This test uses a small source (less than 1 cm in diameter) containing about 200 MBq ^{99m}Tc. The collimator to be tested is fitted and then the detector rotated so that the collimator is vertical and facing outwards. The source is placed a few metres away from the camera so that a very blurred image is formed. The exact distance of the source from the collimator is not important, but the further away the source is placed, the larger the image will be. Ideally, the blurred image should fill a significant part of the field of view. If it does not fill the entire field of view, then several separate images should be acquired with the source repositioned to cover different areas of the collimator each time.

Each image should be acquired on a 256×256 matrix with about 5 million counts and then displayed in a discontinuous colour scale like the one shown in Figure 5.3. If the collimator holes are perfectly parallel, then each image should appear circular like the example in Figure 5.3a. With some collimators, the pattern can have a somewhat hexagonal shape reflecting the distribution of holes, but it should always be symmetric. An irregular pattern like that shown in Figure 5.3b is an indication of badly aligned holes in the collimator. This test is well worth performing as an acceptance test for newly purchased collimators if they are intended to be used for

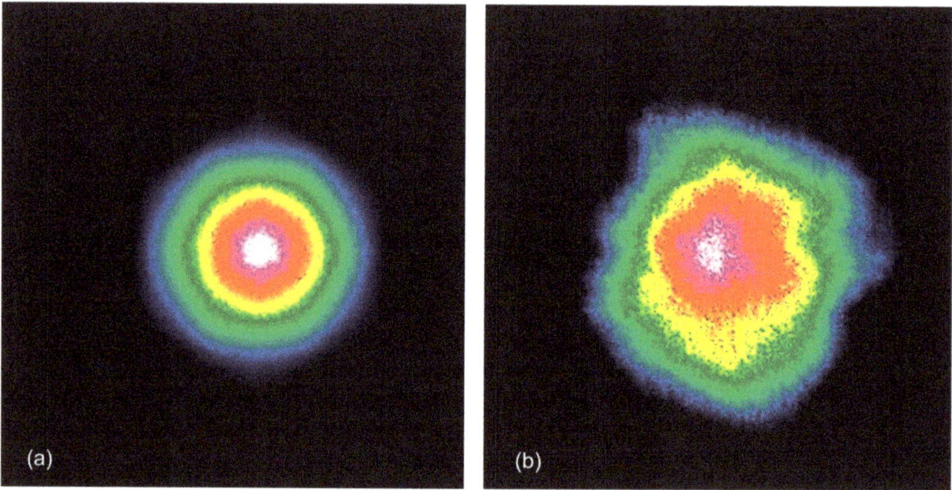

Figure 5.3 Images from a collimator hole angulation test. (a) A good collimator. (b) An unacceptable collimator (reproduced with permission from Lawson (2013))

SPECT. A poor result like that in Figure 5.3b would be grounds for rejecting the collimator.

Protocol for collimator hole angulation check

a) Fit the required collimators.

b) Place a source of about 200 MBq ^{99m}Tc at a distance of a few metres from the collimator.

c) Acquire an image on a 256 × 256 matrix with 5 million counts. If necessary, acquire further images with the source repositioned to cover the full field of view.

d) Examine each image for any deviations from a symmetric appearance.

e) Repeat for each detector and for all collimators.

5.6.2 Frequency of collimator hole angulation testing

The collimator hole angulation test should be performed as part of acceptance testing for any new camera. Thereafter, it only needs to be repeated if collimator damage is suspected.

5.7 Consistency of angular response

Ideally, the uniformity and sensitivity of a detector should not change with angle as it rotates around its orbit. A correctly functioning system should exhibit local and global changes in sensitivity below 1%. Modern systems incorporate sufficient magnetic shielding to maintain this condition with rotation in the Earth's magnetic field; however, there may be local fields which are sufficient to modify the performance of the detector. There may also be other effects, such as changes in thermal gradients within the detector housing which can affect the sensitivity or uniformity. The previous version of this report (IPEM, 2003a) described a method for quantifying the variation in detector response with projection angle suggested by the American Association of Physicists in Medicine. However, this report now recommends a simplified version of the test (AAPM, 1995).

A ^{57}Co sheet source should be firmly and securely fixed to the face of the collimator taking care not to interfere with any air intakes to the camera head which might affect temperature. Check that the source will rotate safely with the detector without hitting anything. Uniform flood images should then be acquired with the detector positioned at four orthogonal positions around its orbit. The sensitivity (determined from the total count rate) should not differ by more than 1% between positions. The uniformity pattern should also not vary with position.

Protocol for consistency of angular response

a) Fit the LEGP collimators.

b) Attach a ^{57}Co flood source securely to the face of one collimator.

c) Move the detector to the 0° position.

d) Acquire a static image using a 64 × 64 matrix with a preset acquisition time sufficient to give at least 5 million counts.

e) Acquire four further static images using the same preset acquisition time but with the detector positioned at 90°, 180°, 270°, and 360°.

f) Determine the total counts in each of the five acquired images. No image should differ from the average counts by more than 1%.

g) Subtract the 0° image from each of the other four images. There should be no significant structure in the uniformity pattern of any of the difference images.

h) Subtract the 270° image from the 90° image. There should be no significant structure in the uniformity pattern of the difference image.

i) For multi-detector systems, repeat for each detector.

5.7.1 Frequency of consistency of angular response testing

The consistency of angular response test should be performed as part of acceptance testing for any new camera. Thereafter, it may be repeated at monthly intervals if required.

5.8 Detector matching

For a multi-detector SPECT system, each detector is usually responsible for acquiring a separate range of projection views which are then combined together in the subsequent reconstruction. Therefore, it is important that the responses of all of the detectors are closely matched. This should be included as part of acceptance testing. There are three important parameters that need to be considered: the relative sensitivity of the detectors; pixel size calibration for each detector; and registration of the digital field of view for all detector heads.

The relative sensitivity of each detector can easily be determined by acquiring a flood image with a ^{57}Co source for the same time with each detector. The count rate obtained for each detector should agree to within better than 10%.

The pixel size for each detector must be precisely known and well matched for each, because calculated attenuation correction algorithms apply a fixed attenuation correction coefficient to the data obtained from all detectors, expressed in units of either cm^{-1} or $pixel^{-1}$. Pixel size may be determined for each detector by acquiring an image of two point sources or two thin lines placed on the collimator face at a fixed distance apart and orthogonal to one axis of the detector. Data should be collected into a fine pixel matrix with a profile generated through both sources to obtain their separation in pixel units. From the separation of the pixels containing the peak count for each source and the known distance between the sources, a calibration factor may then be calculated.

As discussed in Section 5.5.2, the correct registration of Y image coordinates between detectors can be determined as part of the centre of rotation check. A misregistration greater than 2 mm should be remedied owing to the potential for degradation in reconstructed spatial resolution.

5.9 Reconstructed SPECT uniformity

Since the SPECT reconstruction process will amplify any small non-uniformities that lie close to the axis of rotation (Lawson, 2013), it is necessary to check the uniformity of reconstructed images as well as the planar uniformity. Therefore, once planar uniformity has been found to be acceptable as described in Section 5.3, reconstructed SPECT uniformity should be assessed as described in this section.

The phantom required is a cylinder with a diameter of about 20 to 30 cm and a similar length. A Jaszczak phantom without any inserts is suitable, but in-house phantoms can also be improvised from any cylindrical plastic or Perspex container that can be filled with water and then sealed without leaking. The phantom should be filled by adding a few hundred megabecquerels of ^{99m}Tc to the water and then mixed thoroughly. To aid mixing, it is helpful if the phantom is not completely filled initially so that a good sized air bubble remains. Then if the phantom is turned over and over, the bubble gives the water space to move around and mix. Finally, the air bubble can be reduced by topping up the phantom with a syringe of water which can be refluxed a few times to ensure that this is also mixed in. It does not matter if a small air bubble remains as this will float to the top of the phantom. The phantom should be placed on the imaging couch with its axis parallel to the axis of rotation. Ideally, the centre of the phantom should be positioned a few centimetres above the axis of rotation so that any artefacts due to the phantom can be distinguished from those due to rotation. A SPECT acquisition should be performed using a circular orbit and a standard clinical acquisition protocol but with a longer acquisition time to give at least 30 million counts in total over all projections. A 128×128 matrix should be used with at least 60 projections over 360°.

The acquired projections should be reconstructed using filtered back-projection with a high cut-off frequency to give only a small amount of smoothing, and attenuation correction should be applied. Transaxial slices approximately 10 mm thick should be displayed and examined for uniformity.

If a circular orbit is used, then any significant local non-uniformities in planar response will generate 'ring' artefacts in the reconstructed transaxial data, centred on the centre of rotation. The magnitude of the artefacts will diminish in inverse proportion to their distance from the axis of rotation (Lawson, 2013). These artefacts are less visible if non-circular orbits are used because each reconstructed pixel has contributions from more parts of the camera field of view which averages out any non-uniform effect. The rings may be 'hot' or 'cold' and will change from slice to slice, so a full range of slices along the length of the phantom should be examined. It may be necessary to acquire studies with the phantom placed in several different positions along the axial direction in order to cover the full detector field of view.

Figure 5.4 illustrates some ring artefacts caused by poor camera uniformity. The images are transaxial sections reconstructed through a Jaszczak phantom which contains cold rods (in slices 1 to 5) and cold spheres (in slices 9 to 16). However, in addition to these inserts, there are hot rings visible in slices 7 to 12 and cold spots in slices 16 to 20. These are ring artefacts centred on the centre of rotation which happens to coincide with the centre of the phantom in this case. Clearly, these artefacts would be easier to see if the inserts were not present in the phantom and more definitive if the centre of the phantom did not coincide with the centre of rotation. These images were obtained from an old gamma camera without full

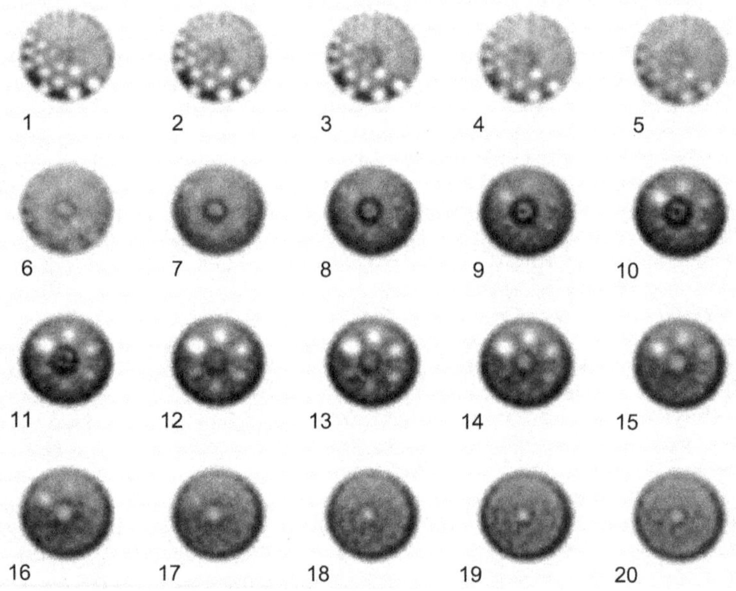

Figure 5.4 Transaxial sections through a Jaszczak phantom. The hot rings in slices 7 to 12 and the cold spots in slices 16 to 20 are due to poor camera uniformity

uniformity correction and it demonstrates that this camera should not be used for SPECT.

Such artefacts are most reliably and sensitively detected through critical visual inspection of the images, but it can also be helpful to generate horizontal and vertical profiles through each transaxial slice. This will enable the count contrast due to any ring artefacts to be determined. The overall flatness of the profile data will also highlight whether adequate attenuation correction has been applied. A concave profile indicates that attenuation has been under-corrected and a convex profile that it has been over-corrected. Incorrect attenuation correction may be due to failure to detect the correct boundary of the phantom, inaccurate pixel size calibration (see Section 5.8), or use of an incorrect attenuation coefficient. For ^{99m}Tc in a water filled phantom, an effective attenuation coefficient of around 0.11 to 0.13 cm^{-1} is usually about right, although this depends on the exact energy window used.

There is no general agreement on how the uniformity of reconstructed slices should be quantified. The American Association of Physicists in Medicine suggest drawing square regions of interest and then calculating the integral uniformity (in a similar way to the NEMA method for planar uniformity – see Section 3.3.4) and the r.m.s noise (similar to the coefficient of variation for planar uniformity – see Section 3.3.4) (AAPM, 1995). If the coefficient of variation is calculated, then it must be recognised that the transaxial image counts have been multiplied by various normalisation factors during reconstruction and so they will not follow a Poisson distribution. Therefore, the correction for Poisson noise described in Section 3.3.4 cannot be applied to reconstructed SPECT data and so the result will be dependent on the number of counts acquired. Alternatively, the International Atomic Energy Agency suggests measuring the contrast of any visible ring artefacts (IAEA, 2009). This report recommends using a simple visual check that there are no visible ring artefacts.

Protocol for reconstructed SPECT uniformity

a) Fit the required collimators.

b) Fill a uniform cylindrical phantom with water and add about 400 MBq ^{99m}Tc. Mix thoroughly.

c) Place the phantom on the patient bed so that the axis of the phantom is parallel to the camera axis of rotation with the centre of the phantom about 3 cm above the axis.

d) Acquire a SPECT study using a circular orbit with 60 views over 360° on a 128×128 matrix. The acquisition time per view should be sufficient to acquire 500 kcount in each view.

e) Reconstruct the data using a filter with a high cut-off frequency and applying attenuation correction. Create 10 mm thick transaxial slices.

f) Inspect the transaxial slices for any sign of ring artefacts near to the centre of rotation. There should be no visible rings.

g) Draw profiles across the transaxial slices and inspect these for flatness as an indication that attenuation correction is correct. Profile counts at the centre of the phantom should be within 10% of counts at the edge.

h) Repeat with all sets of collimators used for SPECT.

5.9.1 Frequency of reconstructed SPECT uniformity testing

Reconstructed SPECT uniformity should form part of acceptance testing and thereafter be repeated periodically at intervals of no more than 1 year. Care must be taken to use the same acquired counts and reconstruction parameters each time so that annual results are comparable with the original acceptance test. Although camera uniformity can change on a much shorter time scale, this should be detected by regular planar uniformity testing (Section 5.3).

5.10 Reconstructed SPECT resolution in air

The resolution of a reconstructed SPECT image depends not only on the resolution of each acquired projection image but also on the fact that these are combined together correctly during reconstruction. Therefore, a test of reconstructed SPECT resolution is a good way of showing that all corrections are working properly. The following procedure is based on that described by the American Association of Physicists in Medicine (AAPM, 1995). The International Atomic Energy Agency recommends a similar method (IAEA, 2009), but NEMA use a completely different technique (NEMA, 2012).

A capillary line source with internal diameter less than 2 mm should be filled with about 30 MBq ^{99m}Tc. Note that this will require a high concentration of technetium eluate. The line should be placed as close as possible to the axis of rotation and parallel to it. The source should be suspended in air beyond the end of the patient couch so that there is no intervening scattering material between the source and the collimator. The detectors should be set at a radius of rotation of 20 cm, measured from the detector face to the axis of rotation. A SPECT study should be acquired using a circular orbit with about 120 projections around 360°. The pixel size should be between 3.0 and 3.5 mm (a 128 × 128 matrix with a zoom factor may be required to achieve this). Each projection image should contain a minimum of 100 kcounts. An additional static planar image of the same line at the same distance from the collimator face should also be acquired with a total of 500 kcounts.

The SPECT study should be reconstructed using filtered back-projection with a ramp filter alone without any smoothing filter applied. Transaxial slices approximately 10 mm thick should be created and horizontal and vertical profiles, one pixel wide, should be drawn through the hottest pixel in each transaxial slice. Spatial resolution should be calculated from the full width at half maximum (FWHM) height of the profile through the line. Because the pixels are rather coarse, it will be necessary to use linear interpolation between pixels to determine the FWHM to within a fraction of a pixel. The result should be compared with the FWHM measured in the same way from the planar image. The FWHM of the reconstructed SPECT spatial resolution should be no more than 10% larger than the equivalent planar spatial resolution. This test should be repeated at two or more further positions along the axis of rotation, in order to sample the full length of the detector face in the Y direction.

It is important for this test that no smoothing filter is applied during reconstruction as a smoothing filter will make the SPECT resolution worse. If the SPECT resolution is more than 10% worse than the planar resolution despite no smoothing filter, then this could be because the electrical and mechanical alignment is incorrect, causing errors in alignment of the projections. These errors should be detected by the COR check described in Section 5.5. However, the reconstruction software used can also affect the result. Some manufacturers provide for the reconstruction process to be optimised either for quantification or for qualitatively enhanced image quality. If this is the case, the quantitative option should be selected for this test in order to ensure a correct comparison. Some software also allows variations to the basic ramp filter which can result in a SPECT resolution which is actually better than the planar resolution. This option should not be used because, although the resolution may appear better, the noise levels in SPECT reconstructions will be higher than necessary, and the linearity of count reconstruction will be poor. If software upgrades are installed, special care should be taken to verify that changes of this sort have not been introduced unknowingly as these can affect clinical results.

Protocol for reconstructed SPECT resolution

a) Fit the required collimators.

b) Place a capillary line source containing about 30 MBq ^{99m}Tc so that it overhangs the end of the patient bed in air. Position the source on the axis of rotation.

c) Acquire a SPECT study using a circular orbit with radius 20 cm. Acquire 120 views over 360° using a 128 × 128 matrix with some zoom. Each view should be long enough to acquire 100 kcounts.

d) Acquire further SPECT studies with the source repositioned along the axis of rotation to sample the full length of the detector.

e) Acquire a static image of the same line source with the detector 20 cm away on a 128 × 128 matrix with 500 kcounts.

f) Reconstruct each of the SPECT studies without any smoothing filter. Create 10 mm thick transaxial slices.

g) Determine the resolution of the reconstructed line in each SPECT study by calculating the FWHM from a profile across the image.

h) Determine the resolution of the planar image by calculating the FWHM from a profile across the image.

i) Compare the SPECT resolution of each study with the planar resolution. The SPECT resolution FWHM should be no more than 10% greater than the planar resolution FWHM.

5.10.1 Frequency of reconstructed SPECT resolution measurement

Reconstructed SPECT resolution should be measured as part of acceptance testing for any new camera. Any changes due to electrical or mechanical effects should be detected by the monthly centre of rotation check. Therefore, the reconstructed SPECT resolution measurement only needs to be repeated after any revisions to the installed SPECT software.

5.11 Total SPECT performance

Total performance of SPECT systems can be assessed for relevant clinical data acquisition and processing protocols using suitable phantoms. Tomographic spatial resolution in the presence of typical levels of scatter may be quantitatively determined using one or more radioactively filled sealed line sources securely fixed within a water-filled (non-radioactive) cylinder. Tangential and radial resolution may then be measured from reconstructed slices in a manner similar to that described in Section 5.10, following the technique specified by NEMA (2012). A line source phantom insert suitable for this assessment can be purchased commercially but it is easy to construct a suitable phantom in-house. Figure 5.5 shows how five lines parallel to the axis of rotation can be created by stretching a length of plastic capillary tubing between two Perspex discs. All five lines can be filled at once by injecting activity into one end of the tubing.

More complex phantoms have been developed to assess multiple aspects of total SPECT performance simultaneously. A widely used example is the Jaszczak phantom which can be used without any inserts to assess tomographic uniformity (Section 5.9) or with rod or sphere inserts to give a semi-quantitative assessment of tomographic spatial resolution. When filling these phantoms, care must be taken to obtain good mixing as already described in Section 5.9. Such phantoms may be used to determine 'benchmark' SPECT performance if the data are acquired under (ideal) high count-density conditions to reflect the limiting performance of the system when it is new. Additionally, a baseline set of data should be acquired under a clinically

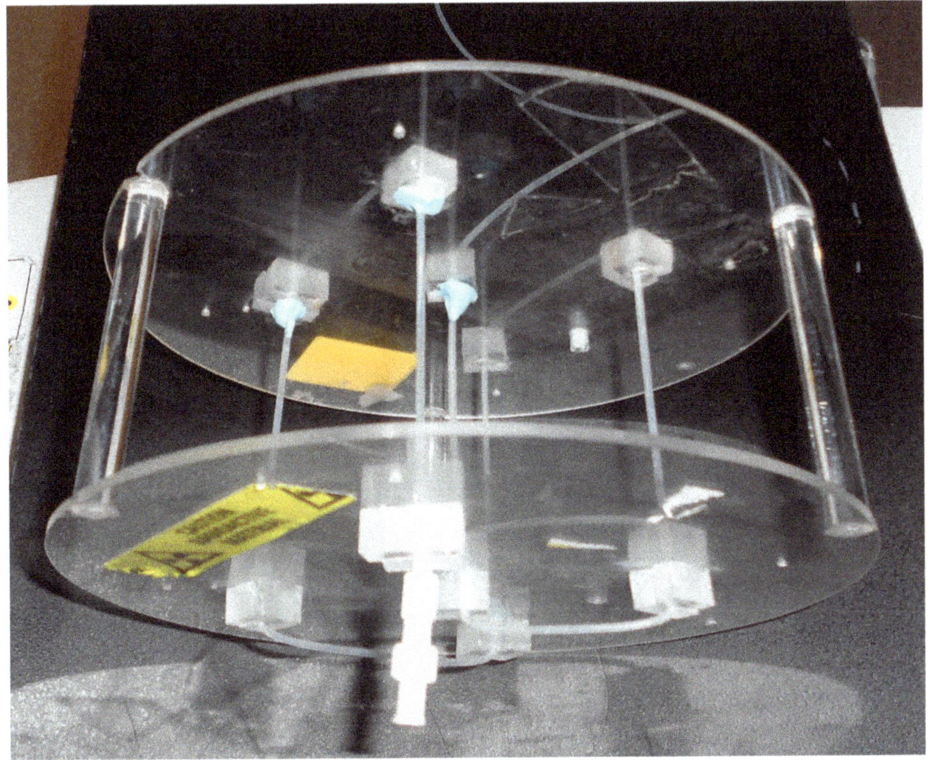

Figure 5.5 An in-house line source suitable for measuring SPECT resolution

realistic protocol to indicate the performance achievable under normal conditions when the system is performing correctly. An objective assessment may then be made of the specific effect achieved by varying individual acquisition and reconstruction parameters within the protocol. This provides a very useful indicator of the SPECT performance achievable with a given system under specific test conditions.

When addressing issues of reconstructed image data quality for any particular clinical application, however, the relationship between this and baseline SPECT is indirect, and to address this issue, specific use may be made of anthropomorphic phantoms. These are designed to mimic the anatomical structure of a specific body region under examination through the accuracy and detail of their construction. When filled using activity concentrations representative of the *in vivo* behaviour of the radiopharmaceutical employed clinically, the data obtained may give a more representative indication of the performance achievable for current or planned applications, although degradation in image quality due to issues such as patient motion, respiration blurring and tracer redistribution is still not included.

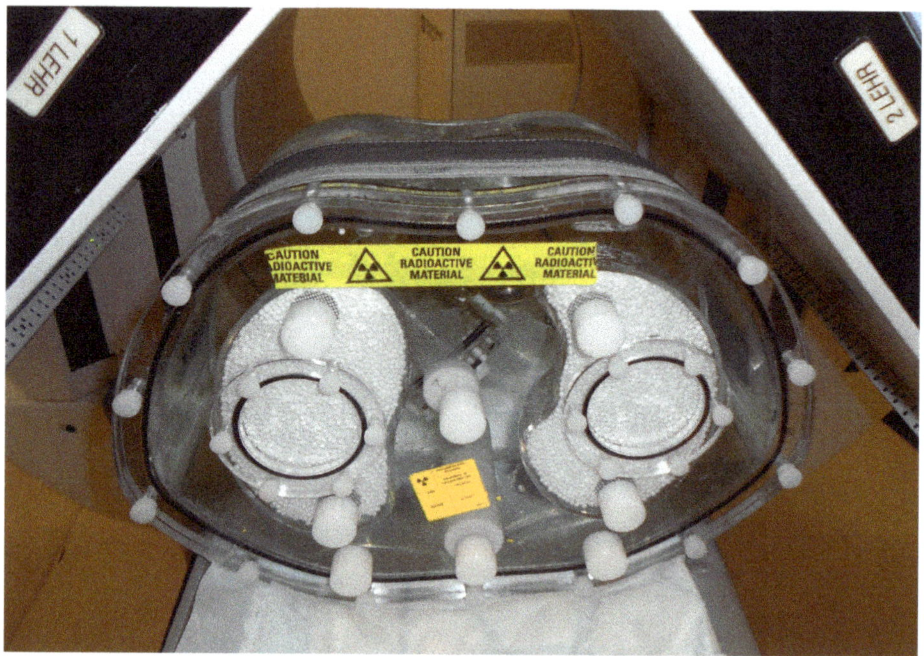

Figure 5.6 A cardiac SPECT phantom positioned under a gamma camera ready for imaging

Anthropomorphic phantoms are commercially available to assess SPECT performance for myocardial perfusion, cerebral perfusion, and neuroreceptor tracers among other applications. Figure 5.6 shows a chest phantom used to simulate myocardial perfusion SPECT. This contains a cardiac insert which can be filled with activity and non-attenuating lung inserts. While the cost of such specialised and often highly complex phantoms is undoubtedly high, their purchase by national, regional, and application-led quality assurance initiatives is often justified. Parameters such as minimum detectable lesion size and minimum detectable contrast ratio may be obtained and investigated further for typical clinical protocols. From these data, together with supporting benchmark data for SPECT performance, a judgement may be advanced in respect of the acceptability (or otherwise) of the performance of a system, and by implication, its 'fitness for purpose' for any specific clinical application. Similarly, such phantoms may be used alongside benchmark performance phantoms to determine the effect of changes in existing data acquisition and processing protocols, and those resulting from system hardware and software modifications.

5.11.1 Frequency of total SPECT performance measurement

The phantom to be used will depend on what is available locally as well as the particular clinical tests performed. However, whichever phantom is used, it is useful to acquire a baseline study at the time of acceptance testing and to repeat this annually using the same conditions.

5.12 SPECT reconstruction software

Very little has been written about the effects of SPECT reconstruction software on the diagnostic accuracy of clinical data, especially where quantification is used. In 2001, the Institute of Physics in Engineering and Medicine Nuclear Medicine Software Working Party conducted an audit of SPECT software within the UK, based on simulated, noise-free, and scatter-free data (Jarritt et al., 2002). The results of this audit highlighted the fact that a number of computer systems do not behave linearly with respect to input count data and that new revisions of software can produce data inconsistent with those acquired using previous versions. Therefore, replacement computer systems may well generate different quantitative responses and hence there will be a need to re-evaluate normal values for clinical data. It is thus essential that users define procedures to test upgrades to system software based upon known test data and that any system replacement is evaluated on the basis of this test data.

5.13 Clinical quality control

A SPECT quality assurance programme should not neglect the processes of data acquisition, analysis, and display for the individual clinical study. It is often in this arena that most significant improvements in clinical image quality can be achieved in addition to improvements in quantitative accuracy. Since the questions posed by each clinical examination are different, and the differentiation of normal and abnormal images is based on different criteria, the QC requirements will often be test-specific. However, the principles outlined here remain foundation stones in this process. Relevant examples have been included in the following sections to illustrate the principles to be followed.

5.13.1 Patient preparation

Several factors remain important in the preparation of the patient. This is true for all imaging studies but is particularly important in SPECT. Patient movement may give rise to artefacts capable of rendering the acquired data uninterpretable. If the radiopharmaceutical has been administered but no diagnostic result can be obtained, this has a clear implication for the radiation protection of the patient. Appropriate counselling should always take place before data acquisition, so that the patient and/or their carer is fully aware of the importance of keeping still throughout the duration of the study. Immobilisation aids and postural support devices should be used as appropriate, and pain relief or sedation should be actively considered if merited.

Scheduling the correct time interval between injection and imaging can also play an important part in obtaining good SPECT images. The preconditions of the tomographic process require that the distribution does not change during the acquisition, thus an imaging time must be chosen to best accommodate this requirement. For example, with stress myocardial perfusion SPECT using ^{201}Tl, it is essential that SPECT data acquisition be completed as soon as possible after injection at peak stress, because the tracer quickly washes out of the myocardium, so that after about 30 min, its distribution within the myocardium is no longer representative of that under stress. On the other hand, myocardial imaging using ^{99m}Tc agents that remain fixed in the myocardium must be imaged with a suitable delay after injection to avoid interference from non-cardiac activity.

5.13.2 Data acquisition protocols

To obtain optimum quality images, it is clearly important that the correct protocols are used for acquisition of the data. Parameters to be considered include choice of collimator, energy window, pixel size, detector orbit, number of projections, acquisition time per projection, and possibly, electrocardiogram (ECG) gating parameters. The values chosen for these parameters will be dependent on the clinical question to be answered. A discussion of the factors that influence these choices for a range of studies can be found in a recent IPEM publication (Lawson, 2013). Every user should ensure that they have optimum acquisition protocols for each clinical situation and that these are applied consistently.

5.13.3 Data checking

When any patient SPECT acquisition is completed, it must be checked for quality before the patient leaves in case there is a problem, such as patient movement, which might require the acquisition to be repeated.

a) Cine display of projections

All SPECT projection data should routinely be reviewed as a cine loop which shows a rotating view of the patient. Inspection of the cine will demonstrate any sudden axial movement of the patient (along the patient couch) during the study, but it is not so good at detecting lateral movement. The use of a horizontal reference line overlying the image aids assessment of isolated organ movement as may be observed for the heart in myocardial perfusion studies. A slow drift in patient movement is not so easy to detect, but if a dual head system is used, slow drift will show up as a sudden jump half way through the study where data from the two heads are joined together. A jump at this point could also be due to detector misalignment, but this should have been detected beforehand by regular COR checks (Section 5.5.2). Any large patient movement will require the study to be repeated but small movements may be correctable with suitable movement correction software. Manufacturers will often supply software especially for this purpose. However, the integrity of such software must always be fully validated before its clinical use, and reliance upon these programs for the routine pre-processing of acquired data cannot be generally advocated. Strategies for the prevention of such motion (see Section 5.13.1) are always preferable to having to take remedial action.

Inspection of the cine loop should also reveal any truncation of count data due to an organ of interest moving outside the detector field of view. A sudden flash of brighter or dimmer images at some parts of the cine is an indication of a variation in acquired counts. This could be due to an intermittent failure of a detector, in which case the data are probably not retrievable. However, with a gated SPECT study, this effect is more likely to be due to ECG irregularities causing beats to be rejected in some views. In this case, the data may be usable if the system has also acquired an ungated study containing all of the beats at the same time.

It is also possible to assess the presence and effect of extraneous sources of activity such as extravasated injection sites and adjacent organs of high uptake within the detector field of view. Sites of activity, such as the bladder, which change significantly during the study, should also be identified, as reconstructed transaxial sections which include such sources of changing activity are prone to significant reconstruction artefacts. This also applies to the acquisition of scattered photons from other patients or inadequately shielded sources within adjacent rooms or corridors. This issue is of particular concern if patients having a positron emission tomography (PET) scan are allowed to walk past a gamma camera room which will not have adequate shielding in the walls to prevent interference from 511 keV gamma rays. Review of projection data will enable such artefacts to be identified before reconstruction.

b) Sinogram display

The 2D sinogram image is generated for a single transaxial image slice, or a small range of slices, by selecting a horizontal profile from each projection image and forming a stack from these 1D profiles for all projections. Figure 5.7 shows an example from a myocardial perfusion SPECT study. Figure 5.7a shows one of 60 acquired projection images with a horizontal line at the level of the heart indicating the location of the sinogram. Figure 5.7b shows the sinogram constructed by stacking the 60 profiles taken at this level, one from each projection view. The horizontal direction of the sinogram therefore represents position across the patient whilst the vertical direction is the projection number, representing rotation around the patient. Thus, individual organs will appear as helical tracks within the sinogram. The resulting image is useful for detecting transverse patient movement but not as sensitive to axial movement. Any sudden transverse movement of the patient (up and down or side to side) will appear as a discontinuity in the sinusoidal path of an organ. In a dual detector system, a jump at the midpoint of the sinogram is indicative of a mismatch of image position between the two detectors. True

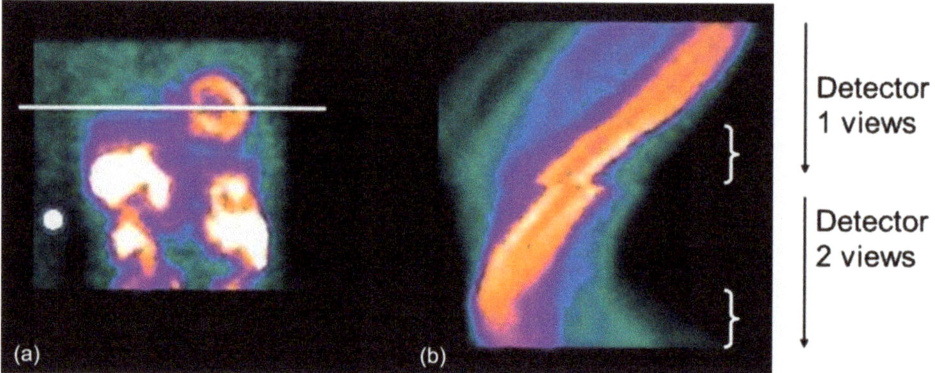

Figure 5.7 Example of a sinogram from a myocardial perfusion SPECT study showing significant patient movement. (a) One out of 60 projection images with a horizontal line showing the location of the sinogram. (b) Sinogram formed from all 60 projections, with braces showing where the patient moved

detector misalignment should have been eliminated by centre of rotation correction, so this is an indication of patient movement.

Figure 5.7b shows a large discontinuity halfway down the sinogram which was caused by significant patient movement. Since this was acquired with a dual detector camera, the first 30 rows of the sinogram were acquired with detector 1 and the remaining 30 rows with detector 2. The patient moved two-thirds of the way through the study and so views 20 to 30 on each detector (indicated by the braces) appear shifted relative to the other views. This is most apparent at the midpoint of the sinogram where view 30 of one detector is placed next to view 1 from the other detector. A sideways jump at this point is a clear indication of patient movement. However, sometimes an apparent discontinuity may just be caused by differences in size of the heart image which can simply be due to different resolutions if the two detectors have been different distances from the heart at this point. As with axial movement (see above), movement correction software may be able to deal with small transverse movements, but large movements, such as that shown in Figure 5.7, may require the acquisition to be repeated.

c) SPECT-CT registration

If hybrid SPECT-CT systems are used, then it is important to check that the SPECT and CT images are correctly aligned (see Chapter 6). Manufacturers usually provide software to enable the two images to be viewed in an overlay mode so that registration can be assessed. Small adjustments may be possible using this software, but severe misregistration (due to patient movement) may require the study to be repeated. However, since this will require a repeat of the CT component, this must be re-justified as it may be that the SPECT study alone will give adequate clinical information. It is also important for data to be reviewed to assess the impact of data truncation on attenuation correction if the patient does not fit entirely within the CT field of view.

5.13.4 Data analysis protocols

Quality control measures are equally applicable to the processes of reconstruction, display, and review of clinical SPECT studies. Wherever quantification of results is required, it is important that the same reconstruction filter is used for all studies. If the filter cut-off frequency is changed, this will affect the reconstructed resolution obtained in the final images. This in turn will modify the magnitude of the partial volume effect and hence can change the contrast of an abnormality which may affect the quantification obtained. This may make it necessary to use one filter (with fixed parameters) for quantification and another for optimal visual interpretation. Where 'reference data sets' are used for data comparison, it is essential to ensure that patient and reference data sets are acquired and processed in the same way. If changes to data analysis methods are implemented, either through improved techniques or software upgrades, the relevance of the reference data set must be reassessed. This is particularly important where gamma camera systems are replaced. The reference data set obtained on a previous camera may no longer be valid if collimation and reconstruction software are different.

Visual interpretation of images is also dependent on the consistent display of the data. This will often require that the reconstructed transaxial slices are reoriented and interpolated to provide standardised transaxial, coronal, and sagittal sections. Particular oblique planes through the reconstructed volume may also be required to improve interpretability of data (e.g. for neurological SPECT or myocardial perfusion SPECT) or for specific circumstances (e.g. dimercaptosuccinic acid (DMSA) SPECT for renal transplants or bone SPECT for misaligned vertebrae). Care must be taken when producing these reoriented slices because there is potential to generate artefactual distortion in the resulting data if there is a mismatch between transaxial slice thickness and within-slice spatial resolution. This can give reoriented slices different appearances in each direction. The same problem can occur if a 1D filter has been applied to profile data during the reconstruction process because this smoothes within each slice but not between slices. It is preferable to apply a 2D filter to individual projections prior to reconstruction or apply a 3D filtering after reconstruction. These requirements are best met by the implementation of standardised protocols at all stages of the SPECT process. Procedures for the evaluation of systems following software and system upgrades are essential.

Since gamma cameras now come with their own computer systems which undertake on-line and on-the-fly corrections for uniformity, sensitivity, and centre of rotation, great care must be taken when the data are reconstructed on a computer system other than that associated with the gamma camera. It is necessary to know whether these corrections have been applied to the raw data before it was transferred or whether a correction has to be applied by the reconstruction computer.

5.14 Summary of measurement protocols

For easy reference, this section contains a summary of the acquisition parameters recommended for each of the measurements described in this chapter. Full details can be found in the relevant sections.

5.14.1 Centre of rotation check

For full details, see Section 5.5.2

Parameter	Value
Collimator	Each set used for SPECT
Source	0.1 mL in 1 mL syringe without needle
Activity	20–50 MBq ^{99m}Tc
Matrix size	128 × 128
Number of views	30 views over 360° for each detector
Time per view	10 s
Reconstruction	Not needed
Analysis	Find centroid of source on each view – Equation 5.1 Fit X values to sinusoid – Equation 5.2 Fit Y values to constant – Equation 5.3
Notes	Repeat for cardiac mode if used
Frequency	Acceptance testing Monthly QC After service or repair

5.14.2 Collimator hole angulation check

For full details, see Section 5.6.1

Parameter	Value
Collimator	Each set used for SPECT
Source	Syringe or sealed tube ~3 m distant
Activity	200 MBq ^{99m}Tc
Matrix size	256 × 256
Acquisition	Static acquisition with 5 million counts
Analysis	Visual inspection for symmetry of image
Frequency	Acceptance testing If collimator damage is suspected

5.14.3 Consistency of angular response

For full details, see Section 5.7

Parameter	Value
Collimator	Low-energy, general purpose
Source	^{57}Co flood source fastened to collimator
Matrix size	64 × 64
Acquisition	Static images at 0°, 90°, 180°, 270°, and 360°
Preset time	Same time for each. Sufficient to acquire 5 million counts
Reconstruction	Not needed
Analysis	Determine counts in each image relative to mean Subtract 0° image from the other four Subtract 270° image from 90° image
Notes	Repeat for each detector
Frequency	Acceptance testing Monthly QC (optional)

5.14.4 Reconstructed SPECT uniformity

For full details, see Section 5.9

Parameter	Value
Collimator	Each set used for SPECT
Source	Uniform cylindrical phantom
Activity	~400 MBq ^{99m}Tc
Matrix size	128 × 128
Number of views	60 views over 360°
Time per view	Long enough to give 500 kcounts
Reconstruction	High filter cut-off. With attenuation correction
Analysis	Inspect slices for ring artefacts and flatness
Frequency	Acceptance testing At least annually

5.14.5 Reconstructed SPECT resolution

For full details, see Section 5.10

Parameter	Value
Collimator	Each set used for SPECT
Source	Capillary line on axis of rotation
Activity	~30 MBq ^{99m}Tc
Matrix size	128 × 128 with zoom
Number of views	120 views over 360°
Time per view	Long enough to give 100 kcounts
Reconstruction	Ramp only if filtered back-projection (FBP), no smoothing filter
Static acquisition	Planar image of same line at same distance
Analysis	Measure FWHM of profile through reconstructed line Compare with FWHM of planar image of line
Frequency	Acceptance testing After SPECT software change

6 Quality control of gamma cameras with CT capability

Mike Avison and David Platten

6.1 Part I: CT

6.1.1 Introduction

It is expected that in-depth image quality and radiation dose tests are carried out by a diagnostic radiology physicist or technologist on an annual basis. More frequent testing should be carried out by nuclear medicine staff. Regulation 31(2) of the Ionising Radiations Regulations 1999 requires that a critical examination be carried out by the installer to ensure that the equipment is operating in a safe manner (IRR, 1999). IPEM Report 107 provides guidance on interpreting this requirement (IPEM, 2012). The critical examination is often carried out by the hospital's diagnostic radiology physicist on behalf of the installer, and combined with acceptance testing to ensure that the system is operating as expected.

IPEM Report 91 provides guidance on which tests to carry out, the testing frequency, and appropriate tolerance levels for a range of diagnostic radiology X-ray equipment, including computed tomography (CT) scanners (IPEM, 2010). Detailed information on how to carry out the CT tests can be found in IPEM Report 32 part III (IPEM, 2003b). These measurements do not replace those daily and weekly tests recommended by the manufacturer, which should be carried out before commencing additional testing. However, it is acceptable to re-use the results of a manufacturer test if it fulfills the requirements of the tests outlined here.

The aim of this section is to describe the tests that are relevant to a CT scanner that forms part of a hybrid nuclear medicine imaging system and would normally be carried out by nuclear medicine staff. The tests assess image quality; there are no direct tests of radiation dose levels. However, the stability of radiation dose can be inferred from the image quality results. A summary of how acquisition and reconstruction parameters affect image quality and patient radiation dose is also provided.

It is important that each test is carried out in a consistent and reproducible way, using the same scan protocol. Changes in acquisition or reconstruction parameters can significantly affect the results. Phantoms should be set up in the same way every time. It is best to centre phantoms in the gantry. The scanner's automatic exposure control, if available, should be switched off for the tests.

6.1.2 Image noise

IPEM Report 91 recommends that noise is measured on a daily to weekly basis (test CT01).

Image noise refers to the spread of pixel values about the mean in an image of an object with uniform density. Noise arises from statistical uncertainties in the measurement process. Changes in image noise can be due to many factors, including the dose received by the detectors, X-ray beam collimation, reconstructed slice thickness, reconstruction filter, reconstruction method, and beam-shaping filter. Figure 6.1a shows a CT image taken through a cylindrical water phantom reconstructed with a 0.5 mm slice thickness. Figure 6.1b was acquired at the same scan settings as Figure 6.1a, but reconstructed to a 3 mm slice thickness, resulting in lower noise.

Noise measurements should be made by scanning a uniform phantom, filled with water or made from a uniform plastic. Noise is quoted as the standard deviation of the pixel values in a region of interest (ROI) placed on the image. A large ROI should be used to avoid local variations due to the structured nature of noise in CT images. This local variation is easy to demonstrate by putting a small ROI on the image and moving it around – variations in the standard deviation will be seen. IPEM Report 91 recommends an ROI diameter that is 40% of the phantom diameter (Figure 6.2).

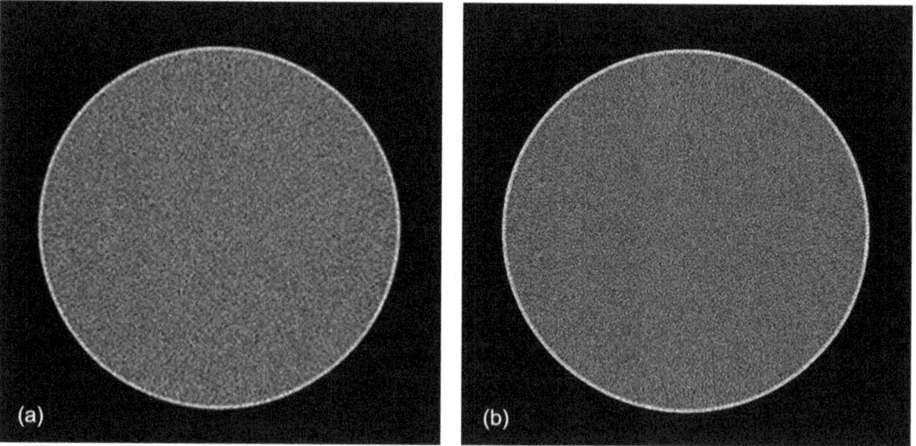

Figure 6.1 Images of a cylindrical water phantom reconstructed with a 0.5 mm (a) and 3 mm (b) slice thickness. The standard deviation of pixel values in a large central region of interest is (a) 167 and (b) 86

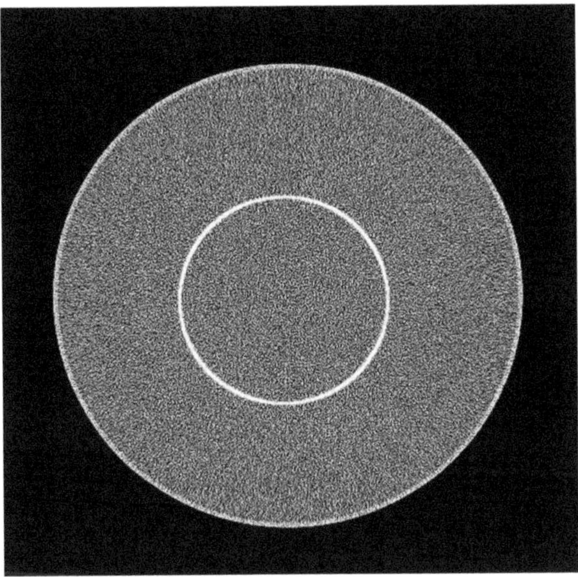

Figure 6.2 Image of a cylindrical water phantom showing a suitably-sized region of interest for noise analysis

The scanner will usually have a phantom supplied by the manufacturer that contains a suitable section for noise measurement. The scan protocol chosen for the test should be clinically relevant.

Acquiring images in a multi-slice axial mode allows variation in noise between the inner and outer slices to be compared. Change in noise in the outer slices is an indirect indicator of changes in the X-ray beam collimation.

The IPEM 91 remedial level for image noise is baseline $\pm$ 10%; the suspension level is baseline $\pm$ 25%. The report also states that, for a multi-slice acquisition, the inter-slice variation in noise should be within 10% of the mean.

Noise level is affected by scan and reconstruction parameters in the following ways:

- inversely proportional to (X-ray tube current)^{½}
- inversely proportional to (scanner rotation time)^{½}
- approximately inversely proportional to X-ray tube voltage
- inversely proportional to (reconstructed slice thickness)^{½}
- dependent on reconstruction filter
- dependent on reconstruction method (filtered back-projection, iterative methods, etc.).

6.1.3 CT number accuracy

IPEM Report 91 recommends that CT number accuracy is tested on a daily to weekly basis (test CT02).

The CT number, measured in Hounsfield Units (HU), of a given material should be constant for a particular X-ray spectrum. The CT number is related to the attenuation coefficient of the material relative to that of water, as shown by Equation 6.1. If the CT images are to be used for attenuation correction, it is important that the CT numbers are correct.

$$CT\,number = \frac{\mu_{tissue} - \mu_{water}}{\mu_{water}} \times 1000 \qquad (6.1)$$

The tests are carried out using a phantom containing a range of material inserts of different densities. Ideally, the materials will include a representation of air, soft tissue, water, and bone. Figure 6.3 shows an image through a suitable phantom. If such a phantom is unavailable, materials in existing phantoms could be used, such as air (outside a phantom), water, and plastic.

The CT number is calculated as the mean of the pixel values in a ROI placed on the material of interest. Care should be taken to ensure that the ROI is entirely contained within the material of interest to avoid background materials from contributing to the mean and skewing the result.

Figure 6.3 Image of a cylindrical phantom containing materials with a range of densities for CT number assessment

CT number values depend on the X-ray spectrum, so the tests should be carried out at the X-ray energies in clinical use. Some scanners use different beam-shaping filters for head and body modes. These will result in different X-ray spectra, and so the tests should be carried out in both modes if these are used clinically.

The IPEM 91 remedial levels for CT number accuracy are baseline $\pm$ 5 HU for water and baseline $\pm$ 10 HU for other materials. The suspension levels are baseline $\pm$ 20 HU for water and baseline $\pm$ 30 HU for other materials.

CT numbers are affected by scan parameters in the following ways:

- dependent on X-ray tube voltage
- dependent on beam-shaping filter.

6.1.4 Radiation dose

IPEM Report 91 recommends that the computed tomography dose index (CTDI) is tested on an annual basis (test CT10).

It is not necessary for this to be tested on a more frequent basis by nuclear medicine staff. A full description of dose in CT is beyond the scope of this report. The reader is directed to the European Commission final report EUR 23464 (Wambersie, 2008) for detailed CT dose information. The two main CT dose descriptors are outlined below.

a) Computed tomography dose index (CTDI)

The computed tomography dose index (CTDI), expressed in mGy, describes the radiation dose from a single rotation of the X-ray tube and detectors. It can be measured in air, and also inside standard 16 cm and 32 cm diameter Perspex phantoms. Measurements of CTDI at the centre and periphery of the Perspex phantoms enable volume CTDI ($CTDI_{vol}$) to be calculated.

b) Dose length product (DLP)

The dose length product (DLP), expressed in mGy.cm, is the other key CT dose descriptor. It describes the total dose from a complete scan, and can be related to radiation risk if the scanned area of the body is known. Factors to convert DLP into effective dose are available in NRPB Report W67 (NRPB, 2005). The effective dose can be useful when comparing the relative risk of CT and radionuclide procedures. DLP is calculated by multiplying the $CTDI_{vol}$ by the scan length.

c) Diagnostic reference levels

Regulation 4(3) of the Ionising Radiation (Medical Exposure) Regulations 2000 requires that diagnostic reference levels (DRLs) are set for each standard radiological investigation (IRMER, 2000). DRLs are intended to represent the typical radiation dose received by a standard-sized patient for a particular examination. For CT examinations, DRLs can be set in terms of typical DLP and $CTDI_{vol}$ values. DRLs should be reviewed on an annual basis; an investigation should be carried out if DRLs are consistently exceeded for a particular examination. IPEM Report 88 provides further information on DRLs (IPEM, 2004).

Public Health England (PHE) publishes recommended national reference doses (NRDs) for a range of CT examinations in terms of $CTDI_{vol}$ and DLP (PHE, 2014). These NRDs can be useful to compare with locally determined DRLs; local values that exceed the NRD may be an indicator of poor radiological practice. The recommended NRDs are updated regularly by PHE following national surveys of CT dose.

Radiation dose is affected by scan parameters in the following ways:

- proportional to X-ray tube current
- proportional to scanner rotation time
- proportional to (X-ray tube voltage)^2
- dependent on the beam-shaping filter
- dependent on the automatic exposure control settings.

6.2 Part II: SPECT-CT image registration

With the advent of X-ray CT equipment built into gamma camera gantries, it has become necessary to consider some additional quality control (QC) tests

- Those related purely to CT quality (Section 6.1).
- The accuracy of registration between the SPECT images and CT images.
- The effectiveness of attenuation correction derived from the CT.

The latter two will be discussed in this section.

There are two things worth considering before deciding on the methodology that will be used to test SPECT-CT registration:

- The possible causes of misregistration.
- What clinical purpose the images will be put to.

The manufacturer of the camera usually supplies tools and an acquisition procedure to test for SPECT-CT alignment. This is likely to be perfectly adequate but should be critically examined before being relied on for routine use. If the manufacturer's test seems inadequate, the camera owner should develop their own test equipment and protocol, which should be quite a simple task, as follows.

6.2.1 Requirements of a SPECT-CT registration test

There are three main possible causes of error

1. Differences in linear scale between the SPECT and CT images.
2. Table moving incorrect distance when changing between SPECT and CT acquisitions.
3. Table flexing when changing between SPECT and CT acquisitions (Figure 6.4).

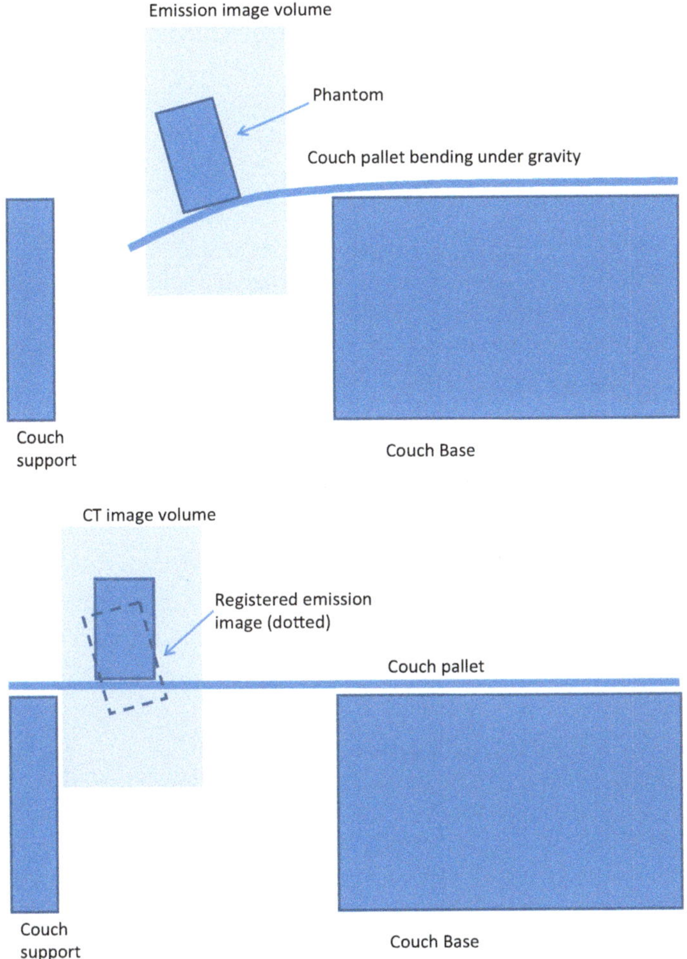

Figure 6.4 Flexing of the couch between SPECT and CT acquisition is one cause of misregistration (exaggerated)

The test should be designed to evaluate misregistration resulting from any of these causes. This imposes the following requirements on the test:

- The registration between SPECT and CT should be evaluated in all three dimensions.
- There should be more than one object in each dimension.
- The objects should be clearly visible with high contrast in both the SPECT and CT data sets.
- The objects should not be on the centre, typically there will be two sources which are displaced either side of the image centre in each dimension by a significant proportion of the field of view, perhaps 20 to 30 cm spacing between them.
- The test should incorporate a weight loading on the table which is representative of the clinical situations which will be encountered – unless it has been firmly established that there can be no significant problems with table flexing.
- Visual inspection of fused data sets will in reality be sufficient to establish co-registration, however mathematical analysis involving the computation of the centroid of each object enables the user to keep a quantitative record.

Table flexing is usually countered by having a support within the gantry which is automatically moved up after the table has passed over it, that is after the table has been extended by approximately 100 cm. This support will usually come into play when the part of the body being scanned is below the shoulders. For heads and necks, the table will not be extended sufficiently for the end of the table to have passed over the support (at least in SPECT acquisition).

When the table is not extended sufficiently to contact the support, light patients will generally be a little too high on the SPECT sequence and heavy patients will be a little too low, because in the SPECT sequence, the vertical position of the patient depends critically on the stiffness of the bed.

Furthermore, when the table is extended well into the gantry, the table may not contact the support unless there is sufficient weight on it. Plainly, it is easy to visually assess whether this is the case. If it is, then the table will have to be loaded with additional weight when performing registration tests. For QC purposes, it may be decided that only acquisitions where the bed is resting on the support during SPECT will be tested, unless there are critical uses of the equipment that demand otherwise.

The tolerance for this test mandated by the manufacturers is usually of the order of 2 to 10 mm (GE Healthcare, 2008; Koninklijke Philips N.V., 2014). As stated earlier, each centre should make a decision as to what tolerance limits are acceptable locally, depending on the purpose to which the scans may be put. For general anatomical localisation, the manufacturer's limits are usually adequate.

a) Suggested design for a locally made registration phantom

If the reader wishes to construct their own phantom for testing registration, a suitable design would be a cylindrical section of stiff foam, 30 cm diameter, containing holes for six 10 mL syringes. Each pair of syringes would be parallel to one of the axes (*x*, *y* and *z*) and be spaced by a known separation such as 20 cm (see Figure 6.5). The syringes are filled with activity (say 20 MBq per mL) and the phantom scanned. When the images are reconstructed, the operator can view a fused data set to confirm that the emission and transmission images of the syringes are aligned.

b) A quick method of assessing misregistration without a purpose-made phantom

A quick and simple alternative to constructing a special phantom or using the manufacturer's one is to fill the Jaszczak phantom and carefully locate it on the table with the phantom axis at approximately 45° to the camera axis. Examining a fused image will give a quick assessment on registration.

6.2.2 Problems caused by misregistration

Misregistration between the SPECT and CT could of course lead to the tracer uptake being attributed to the wrong anatomy. If the CT data is used to perform

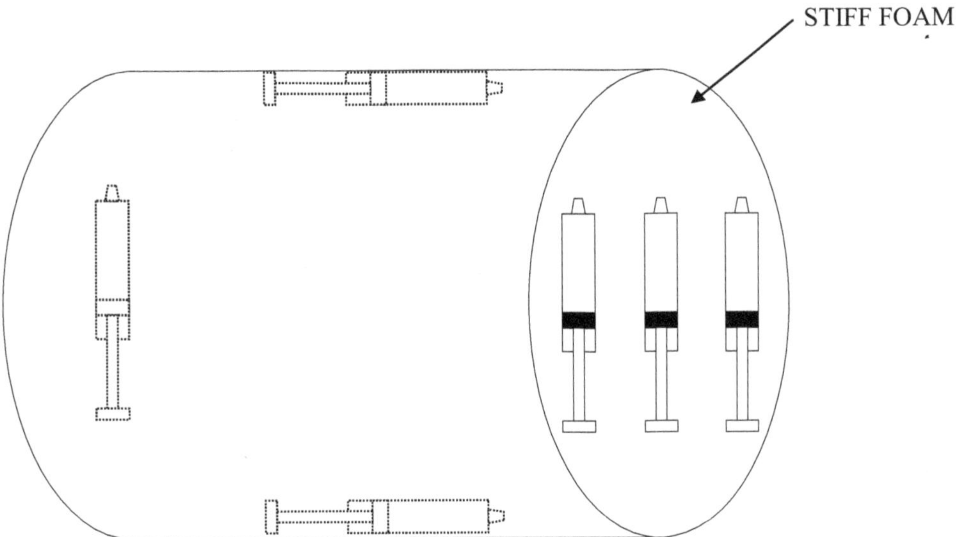

Figure 6.5 Simple design for an alignment phantom

attenuation correction of the SPECT data, misregistration can also lead to errors in this respect (Figure 6.6).

6.2.3 Attenuation correction

Before the availability of transmission images (using an X-ray tube or gamma ray transmission source), the only form of attenuation correction relied on the assumption that the body area being imaged has a uniform linear attenuation coefficient, often referred to as Chang correction (Chang, 1978). Where the mass density of tissues does not vary too much, this is a reasonable assumption. The body area that creates the most problems is the heart; this is because the heart is in close proximity to the lungs and the lungs have a very much lower mass density than adjacent body tissues. In many patients, the inferior wall of the myocardium is below the level of the diaphragm to some extent. Therefore, in these patients, gamma rays from most of the myocardium are attenuated relatively little in the lungs, but those from the inferior wall pass through organs beneath the diaphragm

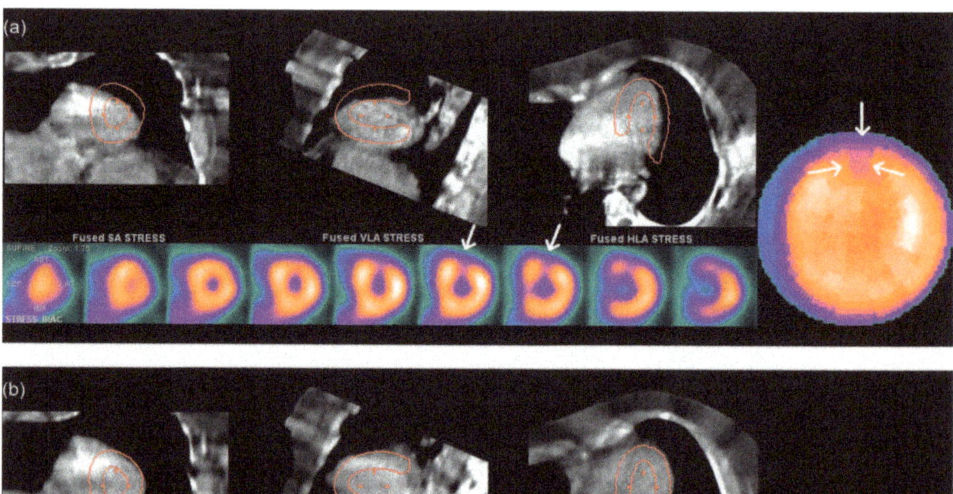

Figure 6.6 Inaccurate attenuation correction caused by misregistration between the transmission and emission data. (a) Prior to correction, an artefact is seen in the anterior wall (arrows). (b) After adjusting the registration between the two data sets

and so are attenuated more. The result is that, in the image, the inferior wall has a lower count density than the other walls. To a lesser extent, other walls of the heart can be similarly affected by overlying breast tissue. This situation can be rectified during reconstruction if a transmission image is available where the voxel values give the linear attenuation coefficient of the tissue concerned.

The end result of this process, if it were operating perfectly, would be to record in each voxel of the SPECT dataset, the counts that would have been recorded in the absence of any attenuating materials. Therefore, the counts in a voxel would be proportional to the activity of material represented by that voxel. The constant of proportionality would then depend on acquisition duration and system sensitivity.

The use of an X-ray transmission map immediately poses problems for designers of the software: The attenuation is measured at the mean energy of photons in the X-ray beam which is significantly lower than the gamma ray energy from most clinical radionuclides. To solve this problem, the computer performing the reconstruction has a file containing a lookup table for attenuation at the X-ray energy to attenuation at the gamma ray energy. Clearly, the validity of this table depends in part on the X-ray beam quality retaining its design specification.

In this process, there is ample scope for errors to creep in. Three of the most likely errors of significant magnitude are

- misregistration (as already discussed),
- a CT data set that is somehow suboptimal, for example, the edge of the patient may be outside the field of view (truncation) leading to underestimation of attenuation. Another possibility is that CT pixel values may be inaccurate due to the patient being positioned non-centrally during the CT acquisition,
- the nature of the algorithm employed by the software writers and the values of constants used by the algorithm, especially the linear attenuation coefficients at X-ray energy and at gamma ray energy.

6.2.4 CT quality for attenuation correction

It is important that the CT scan returns reasonably accurate values of attenuation. The necessary QC for this is generally to scan a phantom containing objects of known and widely varying attenuation, such as air, water-equivalent, and bone-equivalent materials. The manufacturer usually provides a phantom and procedure, including an analysis process which performs these tasks very adequately. In this case, the operator only needs to ensure the test is performed routinely to schedule and the pass/fail advice coming from the analysis program is acted upon. Frequently, the system will not allow clinical scans unless the CT test has been successfully performed within a predetermined time.

6.2.5 QC strategies for attenuation correction

The success of the algorithm will certainly be influenced to some extent by the configuration of activity concentration and attenuating materials in the test object or patient. It is therefore necessary to test the system in a clinically relevant context. Attenuation materials which have a very different effective atomic number from human tissues (for example, iron, copper or lead) should be avoided. The user may wish simply to make an addition to a commercial anthropomorphic phantom. For instance, the user could model the situation where a myocardial perfusion test has to be performed with the patient's arms down, by adding a rod of polytetrafluoroethylene (PTFE) or aluminium to a thorax phantom, either of which has similar density and effective atomic number to bone (Figure 6.7). Alternatively, saline bags taped to the surface could represent breasts. In such cases, mimicking myocardial perfusion scanning, it would be wise to standardise the configuration of the phantom and set action levels based on homogeneity of myocardial counts.

Although not in widespread clinical use at the time of this publication, there is increasing interest in using absolute quantitation of SPECT data to calculate therapy organ dose and monitor tumour response (Seo et al., 2005; Buckley et al., 2007; Zeintl et al., 2010; Ritt et al., 2011; Beauregard et al., 2011; Yoneda et al., 2012). Newer reconstruction algorithms make corrections for scatter and collimator response and are an important step forward in realising these goals.

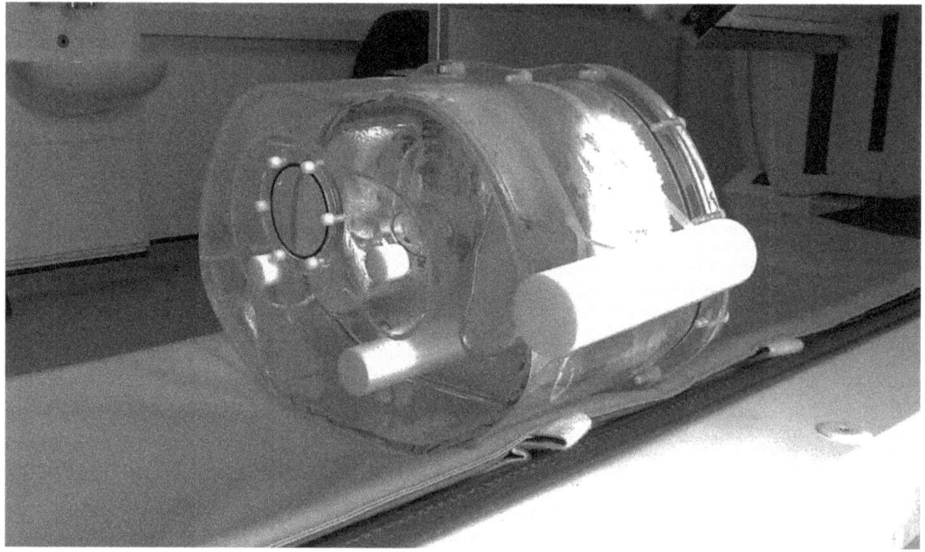

Figure 6.7 Commercial anthropomorphic phantom with the addition of a PTFE rod to simulate a patient's arm

This is a more demanding use of attenuation correction and, in centres embarking on such activities, the design of suitable QC tests will be somewhat different from those just described and will require a phantom with compartments of varying activity concentrations, depths, and sizes.

6.2.6 Recommendation

Owners of SPECT-CT systems should implement tests of CT quality and SPECT-CT registration which best suit the purposes that the camera will be used for. The manufacturers' tests are frequently very suitable but the owner may decide in their case that a locally devised test is required. The frequency of testing should be decided, to some extent, on the historic long-term stability of the camera, but it is recommended that the time between registration tests should not exceed 6 months. Most SPECT-CT cameras force the operator to perform CT quality tests daily.

6.2.7 Clinical QA of transmission scanning

Finally, a word should be said about quality assurance while performing clinical scans. During reconstruction, the registration of the CT to SPECT image should be assessed on every patient, because the patient sometimes moves between the two acquisitions (Section 5.13.1). Modern versions of reconstruction software usually include a utility for both assessing misregistration and correcting for it.

7 Quality control of novel gamma cameras

Ian Armstrong, Natalie Bebbington, Brian Hutton, Therese Soderlund, Penny Thorley and Fred Wickham

7.1 Introduction

A small number of 'novel' gamma cameras which utilise new technology have been introduced by a range of manufacturers in recent years. These novel cameras are mainly small field of view cameras designed for either cardiac or breast imaging and may have a fixed geometry. Cameras include dedicated cardiac cameras with solid state detectors (GE Discovery 530c and Spectrum Dynamics D-SPECT), and cameras with indirect solid state detectors (Digirad ERGO and Digirad Cardius x•act). The ERGO is a large field of view camera used for planar imaging while the Cardius x•act is a dedicated cardiac camera. For cardiac imaging, there is also Siemens IQ•SPECT, an alternative to the solid state cameras, which combines a novel collimator and reconstruction algorithm with a traditional gamma camera. The Oncovision Sentinella 102 portable gamma camera, which utilises a small caesium iodide (CsI) crystal, is also described in this chapter.

7.1.1 Quality control (QC) requirements

The QC requirements of the aforementioned cameras may differ from that of conventional cameras. The standard tests of uniformity and spatial resolution will still be required but customised QC tests may need to be performed for particular systems, and standard QC phantoms often cannot be used due to a small field of view (FOV) and fixed geometry. Collimators are an integral part of some systems so the concept of extrinsic and intrinsic testing does not apply, all tests being extrinsic by default. For those systems only intended to be used for tomography, it is the reconstructed data that reflect the performance of the system. Planar tests may not be either possible or relevant. Performance measures will be affected by the reconstruction parameters and the activity distribution being imaged as well as by the performance of the detector system (Hutton, 2010). Most of these cameras use multiple small cadmium zinc telluride (CZT) or CsI elements (pixels) which may be grouped together into modules or detectors.

7.1.2 QC tests

In general, a smaller number of QC tests may be required due to the limited range of acquisitions available (whole-body imaging may not be performed), and a fixed geometry will preclude Centre of Rotation (COR) testing. Frequency of testing may also be reduced due to the stability of solid state detectors and photodiodes compared to photomultiplier tubes (PMTs).

7.1.3 QC programme

For cameras with solid state detectors, there is a lack of internationally accepted standard procedures both for verifying performance and for monitoring quality. There is also limited advice from the manufacturers in some cases. A QC programme will need to be devised by the user, adapting tests used for conventional cameras to the specific camera system.

7.1.4 Specific camera systems

The following sections deal with QC testing of some of the novel camera systems currently used in the UK. These will aim to inform prospective users of the QC requirements and offer some suggested methods of performing QC.

7.2 GE Discovery 530c

7.2.1 Introduction

The GE Discovery 530c (Figure 7.1) has 19 solid state cadmium zinc telluride (CZT) detectors mounted into a multi-pinhole collimator block. The collimators are

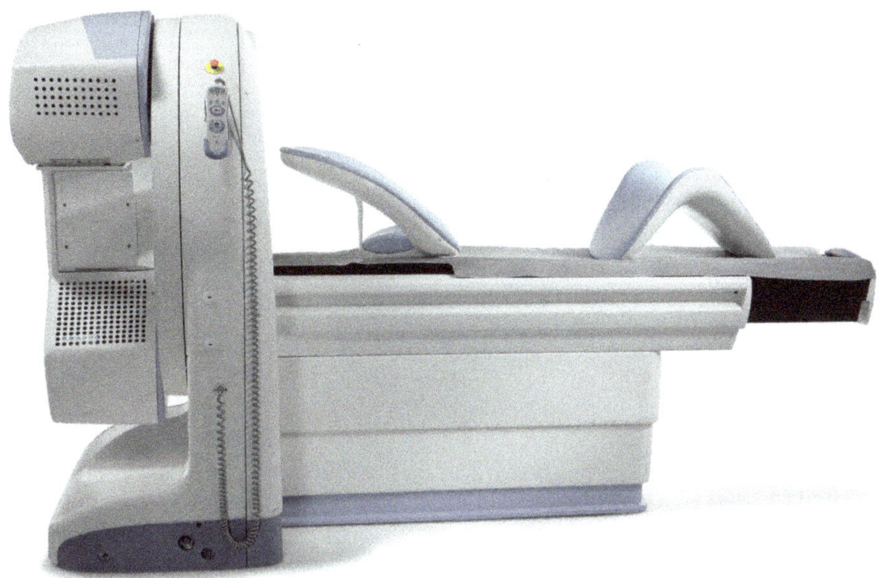

Figure 7.1 GE 530c Cardiac Camera (reproduced with kind permission of GE Healthcare)

positioned to focus on to the heart with a target volume which is a sphere of approximately 19 cm diameter. This allows sufficient 3D sampling for tomographic reconstruction of the heart with stationary detectors. The patient lies (usually supine) on the couch and is moved up as close as possible to the detectors with the heart centred in the field of view. Close positioning is possible as the detector does not move during acquisition. Centring is necessary to ensure the heart is within the target volume as reduced sensitivity and spatial resolution outwith this volume may cause image artefacts.

Each detector head measures 8 × 8 cm and contains 32 × 32 pixelated 5 cm thick CZT elements with a pixel size of 2.46 × 2.46 mm. The distance between the CZT detectors and the pinhole collimator is less than the distance between the collimator and the heart, which minimises the cardiac image to preserve resolution which is stated to be 5.4 mm at 15 cm with scatter compared to 9.4 mm for state-of-the-art conventional SPECT. CZT detectors have enhanced energy resolution at $<6\%$ and show a linear response at high count rates. Sensitivity for the 530c is three to four times higher than for conventional SPECT. This enables cardiac imaging to be performed using much lower acquisition times compared to a conventional gamma camera (typically 5 min compared to 15 min for a 500 MBq dose). Image reconstruction is performed using an ordered subsets expectation maximisation (OS-EM) algorithm to produce reconstructed images comparable to those for conventional SPECT imaging. Raw image data cannot be viewed and all data, including QC images and quantitative values, are dependent on the reconstruction parameters used.

7.2.2 Daily QC

Daily QC is performed using a ^{57}Co flood source (740 MBq and active dimensions of approximately 250 × 250 mm) and a dedicated source holder as shown in Figure 7.2. The source must be repositioned using slots in the source holder for a total of three separate acquisitions taking a total of about 10 minutes with a new flood. Proprietary software then analyses the data and gives an overall pass/fail and clinical protocols are prohibited if the QC fails. Uniformity, energy peaking, resolution, and the number of 'bad' pixels are calculated for each of the 19 detectors, and limits are defined by the manufacturer. Uniformity images for each of the detectors can be visually assessed and exported to enable further assessment with user software if more than the single uniformity value calculated by the system software is required. However, the use or relevance of further uniformity analysis is not known and values would not be comparable to those from conventional cameras. Bad pixels (detector elements) are defined as those where the energy peak or sensitivity is outwith a specific range. There are a total of nearly 20 000 pixels (1024 per detector head) and an upper limit of 1% for allowed bad pixels. Daily QC temporarily deactivates bad pixels.

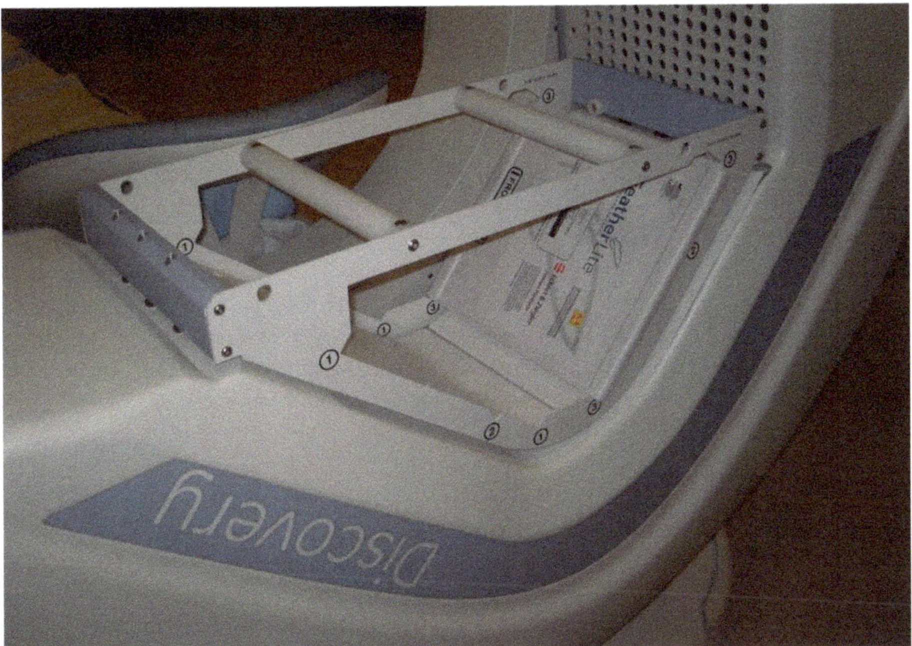

Figure 7.2 Source positioning for daily QC

7.2.3 Periodic QC

Periodic QC is performed once every 2 weeks using the same acquisition jig and takes roughly 40 min. The energy maps of all of the detectors are updated and temporarily deactivated bad pixels are recalibrated. Any pixels which are still 'bad' are then assigned as permanent bad pixels.

Daily QC data can be extracted to a Microsoft Excel file for further trend analysis. Recommended flood life is 2 years and recalibration of the system by GE is required when a new higher activity flood is used.

No other QC is recommended by the manufacturer, so acceptance tests, phantoms, and a QC programme have, by necessity, been devised in-house. The two centres in the UK currently with a GE 530c, Sheffield and Leeds, adopted very similar acceptance tests as published by the Sheffield group (Hillel et al., 2011).

7.2.4 Acceptance testing

a) 3D Spatial resolution

This is measured using a ^{99m}Tc thin line source positioned in the axial and transaxial planes both at the centre and periphery of the imaging field. A capillary held in a jig

(Hillel et al., 2011) can be used or a long line mounted in a phantom (Leeds in-house phantom) to give reproducible positioning. Full width at half maximum (FWHM) (in air) can then be calculated from profiles through the line image, reconstructing using parameters used for clinical imaging.

b) 3D Uniformity

The camera target volume is about 19 cm in diameter so a spherical phantom such as ^{99m}Tc in a water filled Perspex sphere with a diameter slightly smaller than this must be used rather than a Jaszczak or similar phantom. A 16 cm diameter sphere containing 50 MBq ^{99m}Tc in water, centred within the field of view and imaged for 5 min gives good image quality. Reconstruction using standard clinical parameters allows visual assessment, and quantitative uniformity values can be obtained following Chang attenuation correction, as for a conventional SPECT camera. Uniformity values are comparable to those from a conventional camera.

c) Count-rate performance

This can be assessed using the same phantom as used for the 3D uniformity. Repeated measurements (8–10) with activity varying from 50 to 1000 MBq of ^{99m}Tc in water, or measurements as 1000 MBq decays, show a linear response up to about 600 kcps. Calculation is performed as for a conventional camera (Section 3.6).

d) Anthropomorphic phantom

Myocardial perfusion images obtained with the 530c appear different from those obtained by conventional SPECT. Improved resolution and sensitivity result in improved image quality, and attenuation artefacts are different. Anterior attenuation artefacts are reduced as the detectors view the heart from a range of differing views, some of which are not through attenuating breast tissue, but inferior attenuation artefacts are still seen. Further inferior artefacts can also be introduced if the heart is not centred within the focal area. In clinical practice, the heart must be in the same position in the field of view at both stress and rest (within 1–2 cm) to avoid artefactual differences in uptake, particularly in the inferior wall, which can mimic a reversible perfusion pattern. Software is available during acquisition to ensure adequate positioning. Imaging with an anthropomorphic phantom at defined offsets can show this effect and can give an indication of the image quality expected clinically.

7.2.5 Frequency of testing

The 530c has only been in clinical use for 2–3 years so there are no data on long-term stability. It is thought sensible to perform testing at 6 monthly intervals, repeating some or all of the acceptance tests. It may be possible to reduce the frequency of testing once sufficient data have been obtained to ascertain any fall in parameters and what the variation is likely to be. Analysis of 30 million count flood acquisitions acquired at monthly intervals has not been shown to provide any additional useful data and the camera has been stable over the 30 months since installation and no problems with the detectors have been seen.

7.3 Spectrum Dynamics D-SPECT

7.3.1 Introduction

The D-SPECT system (Spectrum Dynamics, Caesaria, Israel) was designed specifically for use in nuclear cardiology (Figure 7.3). The system has a novel design involving nine multi-element CZT detectors, each rotating on its own axis. Each detector consists of an array of 2.46 mm square detector elements (pixels) and extends for 16 cm in the axial direction and 4 cm transaxially. Each detector can be programmed to rotate up to 110° in up to 120 angular increments; the whole set of detectors relocates to a second intermediate position, rotating in the reverse direction, to complete acquisition. Rotation can be fast in a single position (e.g. 3 s) facilitating fast dynamic acquisition. A feature of the system is that the detectors can be programmed to spend additional time acquiring from a selected sub-volume that encloses the heart, thus increasing the number of acquired counts from this region.

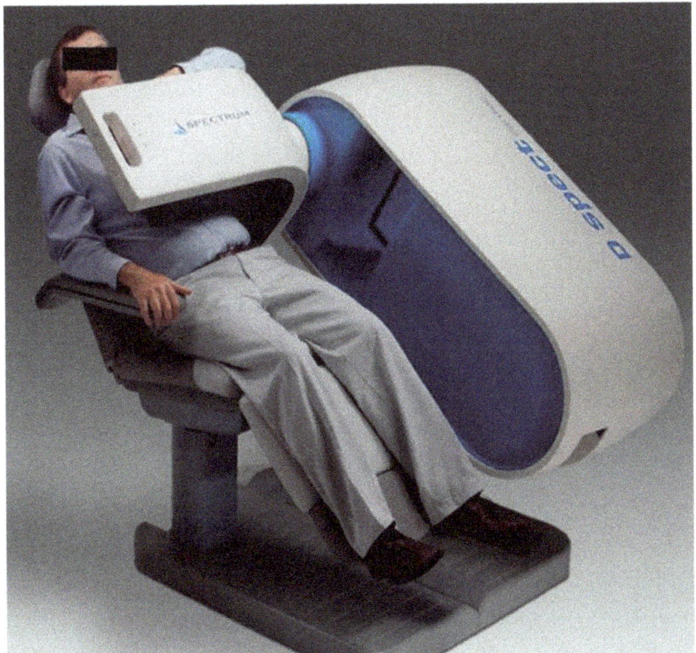

Figure 7.3 D-SPECT system with a patient in semi-upright position (reproduced with kind permission of Spectrum Dynamics)

This, combined with use of wide-angle Tungsten collimators (with apertures matching the CZT elements), provides high system sensitivity. The acquired data are reconstructed using an iterative OS-EM algorithm that includes resolution modelling to produce reconstructed data comparable to that from conventional SPECT. The primary difference from a standard SPECT is the variable angular sampling over the field-of-view, with greater than 180° sampling in the sub-volume close to the detectors that includes the heart but under-sampling in the more distant sub-volume.

Energy resolution can vary for different pixels so any small drift in gain can therefore theoretically lead to inter-pixel variability in observed sensitivity. For this reason, a conservatively wide energy window is used in routine practice. The system calibration is energy-dependent and has to be established for each radionuclide used. Sensitivity correction maps are used similar to those adopted for standard gamma cameras; there is no need for linearity correction due to the pixelated detectors. An important advantage of CZT is that the intrinsic resolution is dependent on pixel size but independent of energy; e.g. thallium-201 images have similar quality to technetium-99m images. Clearly, the system mechanical calibration is important in order to ensure that no degradation in resolution is introduced. There is no centre of rotation as defined for a rotating camera; instead, the spatial and angular positions of the detector have to be calibrated so that the detector locations are known exactly at all times.

The superior energy resolution of CZT permits discrimination of energy spectra for different radionuclides (e.g. the relatively close photopeaks for iodine-123 and technetium-99m). Correction of cross-talk in dual radionuclide studies is somewhat complicated by the presence of a long tail in the energy spectrum below each photopeak due to incomplete charge collection and this must be accounted for (Kacperski et al., 2011).

7.3.2 Daily QC

The daily QC tests verify that the overall system is operating to a satisfactory level; this includes verification of mechanical calibration and detector uniformity as well as the energy resolution. The suppliers prescribe a simple daily QC procedure. This involves mounting a cobalt-57 rod source by attaching a jig to a fixed point on the camera gantry (Figure 7.4). Data are acquired for 120 angular directions covering the full angular span in each of the two detector positions; acquisition takes 5–10 min depending on source strength. Results are presented in a comprehensive single display, which illustrates detector uniformity (visually and quantitatively), an energy spectrum from which energy resolution and peaking are verified, a check that source location does not deviate beyond calibration limits (detector registration) and verification of system sensitivity. A 'traffic light' system (red, amber, green) indicates suitability for clinical use, alerting the user if any parameters are outside the

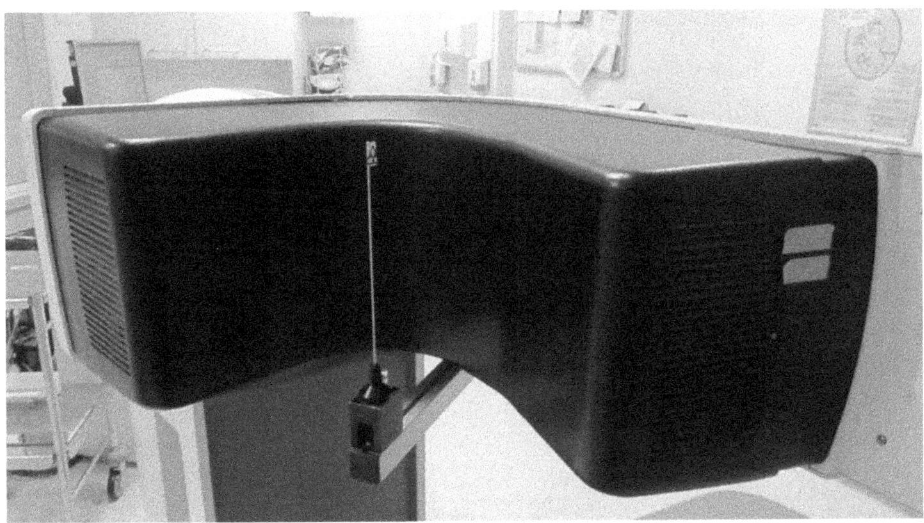

Figure 7.4 Set-up for performing daily QC. A jig is mounted on the camera gantry holding a ^{57}Co rod source in a fixed position

tolerated limits. Quantitative uniformity is reported as global and regional homogeneity indices; these are similar to NEMA figures but represent 100%–NEMA figures for integral and differential uniformity, respectively. Note, however, that uniformity is less critical than for a standard SPECT system as a focal defect does not give rise to a ring artefact.

7.3.3 Periodic QC

Apart from the daily QC, no other user QC is defined by the manufacturer. Regular system calibrations are, however, performed by service personnel. The system's mechanical calibration appears to be very stable, however, calibration of the 'home' position of detectors is checked and recalibrated periodically. The system's energy calibration is subject to change (aging of detector elements, changes in operating environment) and a full recalibration is performed biannually (as well as after software upgrade or major repairs). This is a fairly time-consuming process as it requires a high count acquisition so that meaningful energy spectra can be examined for each pixel; signal gain is adjusted to align the pixel spectra as well as a multiplicative correction to account for any net variation in pixel sensitivity. As mentioned earlier, correction tables are required for each radionuclide used as the adjustments vary with energy. Part of the calibration process also checks for faulty pixels whose signals fall outside acceptable limits or whose response varies significantly with time. These pixels are eliminated with negligible effect on the overall image quality, provided the number of faulty pixels is small. In practice, typically less than 1% of pixels are found to be problematic.

7.3.4 Acceptance testing

As with other non-standard SPECT systems, there is a lack of internationally accepted standard procedures for verifying performance, and acceptance tests currently have to be devised in-house by the user. A comprehensive description of D-SPECT system performance evaluation has, however, been published by Erlandsson et al. (2009), including descriptions and results from count rate, sensitivity, and spatial resolution measurements. Overall, verification of specifications requires a series of tests and control of detector positioning with software that is not normally available to the end user and, as such, requires assistance from service personnel. Operators may wish to verify overall performance using a thorax phantom with cardiac insert.

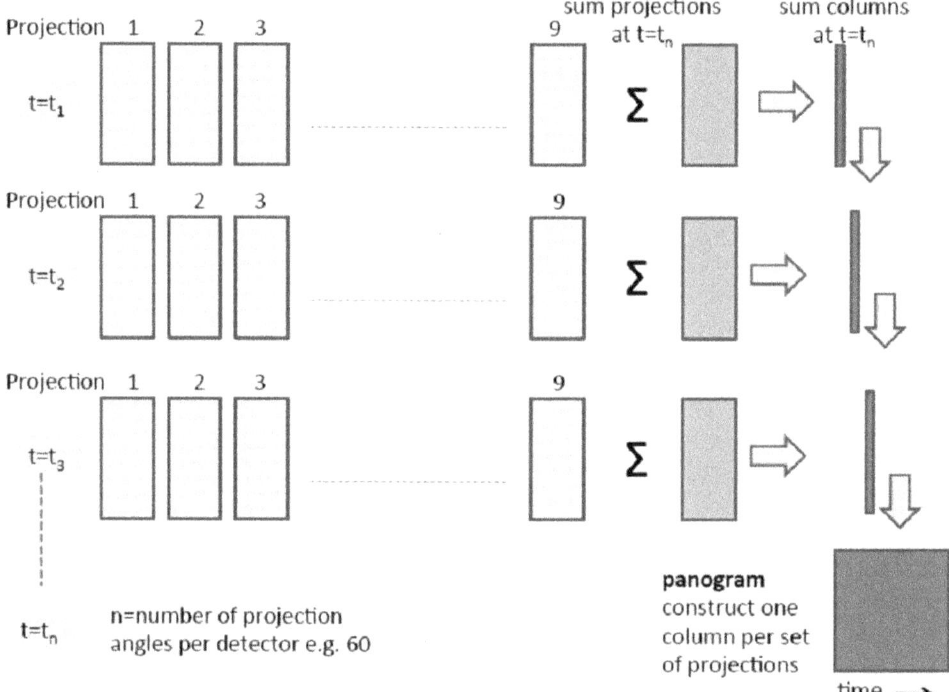

Figure 7.5 The formation of a panogram. All columns from projections acquired from the nine detectors at a particular time are summed to provide a single column in the panogram. When all detectors rotate to the next angle, a further summed column is included. The panogram therefore reflects the change in axial distribution of acquired counts over time, which is sensitive to motion in the axial direction

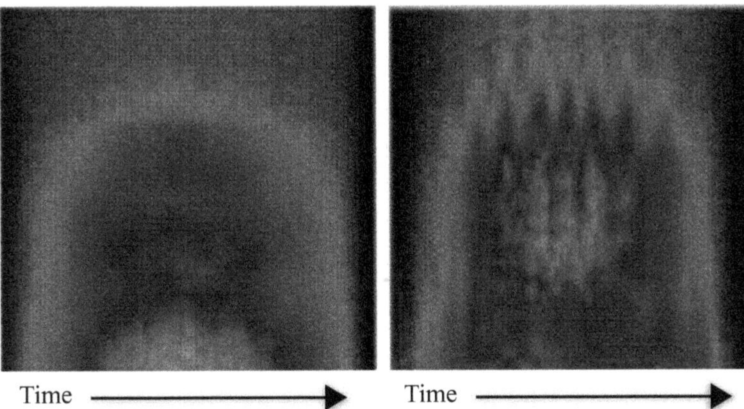

Figure 7.6 Panograms obtained for a good quality scan with no motion artefacts (left) and a scan acquired from a patient experiencing shortness of breath during image acquisition, giving rise to the zigzag appearance of the panogram (right). Note that the y-axis represents distance in the axial direction

7.3.5 QA of image quality during acquisition

The field-of-view during planar acquisition is so small as to make it difficult to clearly identify tracer distribution. The operator is therefore more reliant on reconstructed data when positioning the patient using a short 'scout' scan. However, the operator is also presented with several displays on completion of acquisition to check that data acquisition is complete and to confirm that there were no detector failures and minimal motion. A sinogram is displayed which indicates whether there was any detector failure during acquisition or gross movement. Unlike a conventional SPECT system, motion effects on the sinogram may be difficult to detect as multiple lines of the sinogram are acquired simultaneously and therefore the normally identifiable data displacement is less apparent. A rotating planar cine formed from the multiple detectors is displayed, which is useful in detecting motion, although the planar image quality is poor due to the variable detector–source distances, the poor resolution of the collimator and the variable field-of-view during rotation. Probably most useful is a unique display called a panogram; this is formed by summing counts for all detector columns acquired at a point in time and then displaying the variation in these counts as the detectors rotate (Figure 7.5). Typically, the central portion of the display illustrates increased count density as most detectors will be directed to the heart at mid-acquisition. The display does not provide useful information on the actual activity distribution but is sensitive to motion, particularly in the axial direction (Figure 7.6).

7.4 Digirad ERGO

7.4.1 Introduction

The Digirad ERGO (Figure 7.7) is a general purpose single-detector mobile gamma camera available with a variety of interchangeable collimators. The detector is constructed using a pixelated array of many individual detector elements. Each detector element is a CsI scintillation crystal with a silicon photodiode (SiPD) attached.

The ERGO uses parallel hole collimators, like a conventional gamma camera. Consequently, the methods for performing all extrinsic planar quality control tests: uniformity, spatial resolution, linearity, and sensitivity, are the same as those used on a conventional gamma camera with a collimator. The schedule of routine QC should be the same as that used in conventional cameras.

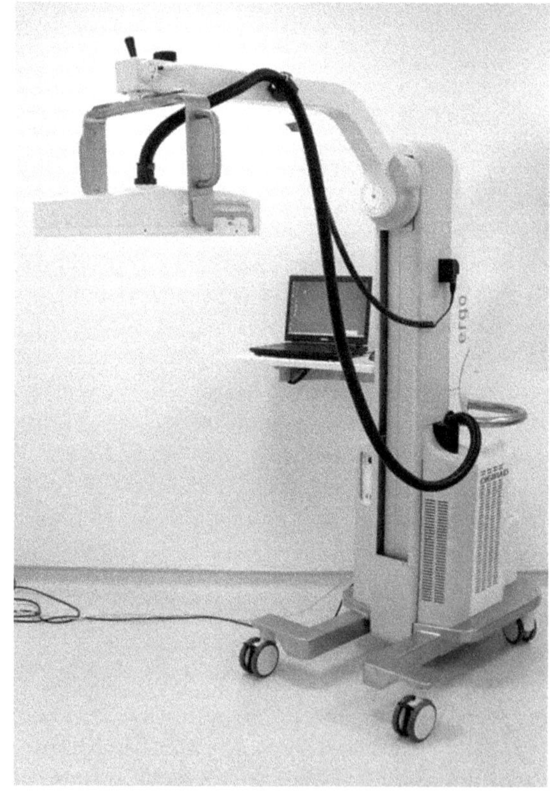

Figure 7.7 Digirad ERGO mobile camera

Measurement techniques on the ERGO which differ from conventional cameras are intrinsic spatial resolution, intrinsic uniformity, and count rate capability.

7.4.2 Intrinsic spatial resolution

Unlike a conventional gamma camera, where the position of an incident gamma ray is determined by Anger arithmetic, the pixelated detector array on the ERGO will determine the gamma-ray position from the individual detector element that the gamma ray deposits its energy in. As a consequence, the limiting factor of the intrinsic spatial resolution is the physical size of the detector elements within the array and so there is little point in measuring the intrinsic spatial resolution. In the ERGO, the size of the each detector elements is 3.0 mm × 3.0 mm.

7.4.3 Intrinsic flood image

The ERGO is not normally intended to be used without a collimator and does not use a uniformity correction map when used intrinsically. Consequently, an intrinsic flood image carried out with a point source at five times the field of view size is very unlikely to produce a uniform image. The uniformity is only correct when collimators are attached, using the specific collimator uniformity correction map.

Despite this apparent limitation of intrinsic imaging capability, an alternative technique may be performed that is intended to assess and monitor the stability of each individual detector element in the system. The ERGO is configured such that a single pixel value in the image matrix represents the number of events detected by a single detector element and therefore each detector element can be assessed from the image. The number of counts in each pixel should be approximately 10 000 to reduce the statistical fluctuations

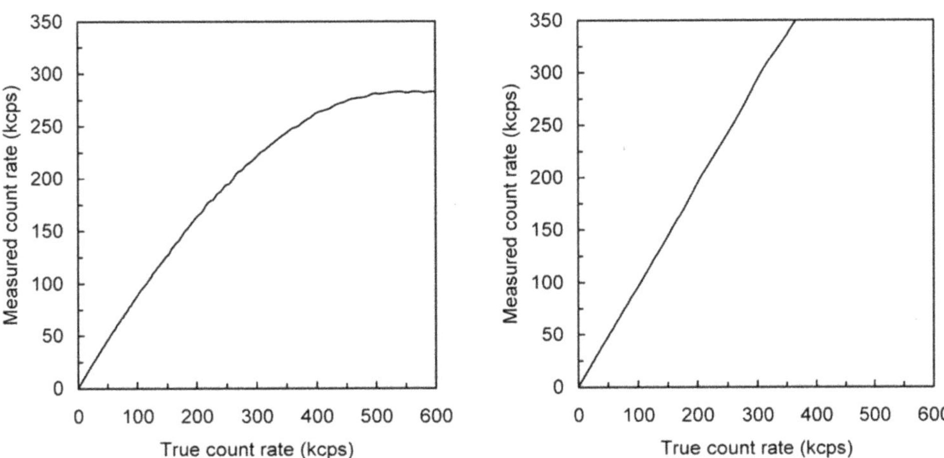

Figure 7.8 Left: Count rate response of a conventional gamma camera (Siemens Symbia). Right: The count rate response of the Digirad ERGO

to 1.0%. The total number of counts in the image that are required is therefore dependent on the total number of detector elements. The technique takes advantage of the linear count rate response at very high count rates (350 kcps) of the ERGO detector which allows for a very high count image to be acquired giving very low statistical fluctuations in the image – see Figure 7.8. The ERGO has 120 × 96 detector elements, which are within a native image matrix of 128 × 128 pixels. To obtain 10 000 counts per pixel, an image requires 115 million counts to be acquired. This may appear impractical in terms of time required to acquire the image but it is acceptable to perform the acquisition with a count rate of 100 000 to 200 000 counts per second with no degradation on the camera performance. Given the low statistical fluctuations in each image pixel, changes in pixel values over time can be used to demonstrate the degree of variation of the detector element response. It should be noted that this test is not one that is suggested by the manufacturer and instead has been developed, including the analysis software, in-house.

Practical measurement

a) The most important factor to consider is the reproducibility of the set-up of the source and detector. In the example used here shown in Figure 7.9, the source to detector distance is set by raising the detector as high as possible on the camera gantry with the detector pointing to the floor. The detector surface should be horizontal and this should be checked with a spirit level. A template of the detector is aligned on the floor using a plumb line suspended from the detector and shows the point to place the source.

b) Place a ^{99m}Tc point source so that it is aligned with the centre of the field of view. The level of activity should be between 100 and 200 MBq of ^{99m}Tc.

c) Acquire an image that gives 10 000 counts per pixel.

Processing of these images requires development of in-house processing algorithms and so processing techniques are up to the individual user. Once acquired, the static images from subsequent acquisitions can be combined into a dynamic image array. Using multiple images allows a standard deviation image to be created, where each pixel value is the standard deviation of the counts of the corresponding pixel from each image.

7.4.4 Count rate

The count rate response of the ERGO is linear over a much greater range than that seen in conventional gamma camera detectors. The manufacturer's specification states that the maximum count rate of the system is in excess of 5 million counts per second. As such, it is perfectly reasonable to assume that count rate losses will not occur in this camera for routine clinical imaging.

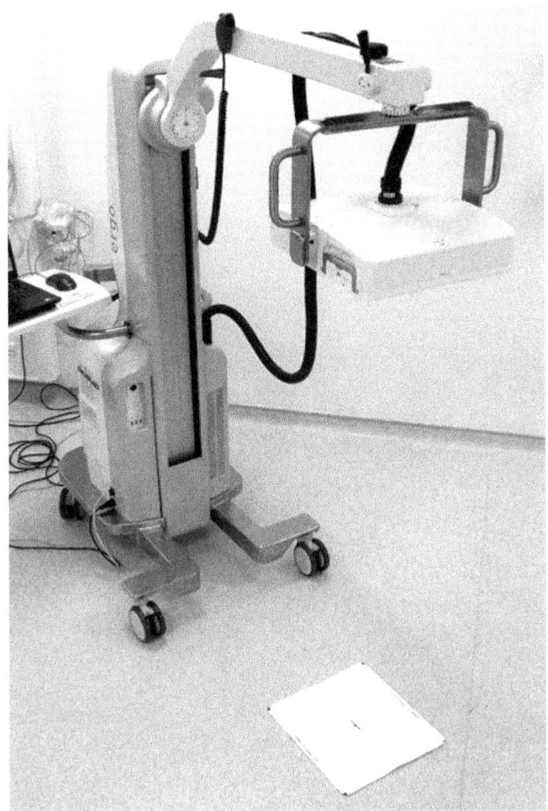

Figure 7.9 Experimental set-up for intrinsic QC image acquisition for the Digirad ERGO mobile gamma camera. A detector template is shown on the floor

7.5 Digirad Cardius x•act

The Digirad Cardius x•act (Figure 7.10) is a triple-detector dedicated cardiac SPECT system. The camera only uses converging collimators that give a magnified view of the heart during the image acquisition. Like the ERGO, the detector is constructed using a pixelated array of many individual CsI scintillators with a silicon photodiode (SiPD). The size of the crystals is 6.1 × 6.1 mm. The system has a very low dose X-ray system that is used for performing attenuation correction for myocardial studies. During the SPECT and CT acquisition, the patient chair rotates to give a tomographic image acquisition.

The collimators cannot be removed by the user on the Cardius x•act and so all image quality tests are performed with the collimators in place. It is therefore not possible to measure intrinsic spatial resolution or uniformity. As with the Digirad ERGO,

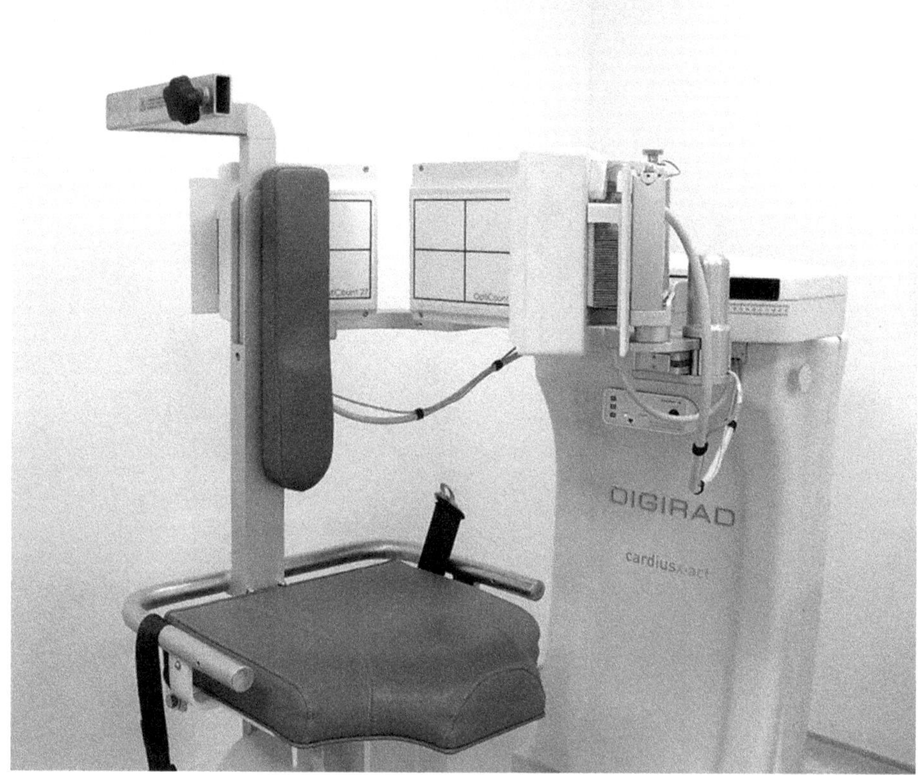

Figure 7.10 Digirad Cardius x•act dedicated cardiac camera

the intrinsic spatial resolution is defined by the size of the individual detector elements. The techniques for planar tests used to measure spatial resolution and sensitivity are the same as those on conventional cameras.

As a result of the positions of the three detectors, it is not possible to get a standard sized ^{57}Co flood source close to an individual detector and so the system uniformity is measured using a tank source filled with ^{99m}Tc that is held against the detector using a special mount (Figure 7.11).

A uniform SPECT phantom can be acquired and processed using the same methods as those described in Chapter 5 of this report on SPECT quality control (Section 5.9). It is also necessary to acquire a blank transmission scan each day using the X-ray system before scanning.

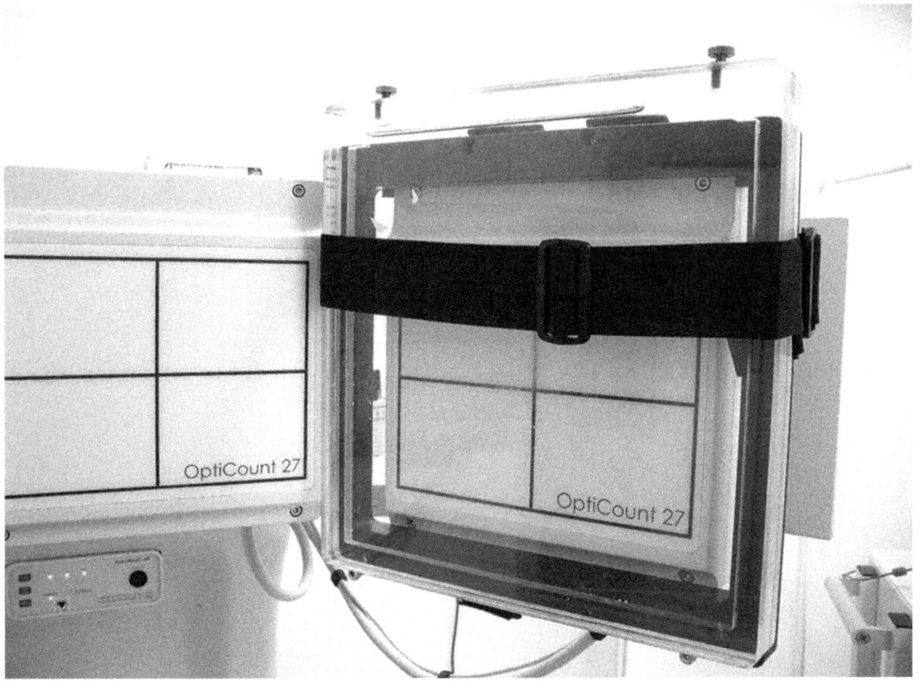

Figure 7.11 System uniformity acquisition using a ^{99m}Tc tank source mounted against a detector of the Cardius x•act

7.6 Sentinella 102

The Sentinella (Figure 7.12) is a portable gamma camera which can be used in an operating theatre. It is mainly used in sentinel node operations. It is also used in brain perfusion imaging for establishing brain death.

The detector of the Sentinella 102 consists of a sodium-doped caesium iodide crystal with 40×40 mm^2 field of view and a position-sensitive photomultiplier. It is supplied with two pinhole collimators with pinhole diameters of 4 mm and 2.5 mm. The detector and collimator have a total mass of around 1 kg and can be mounted on the articulated arm of a cart which has two screens and a processing unit. The system also includes a laser positioning system and a gamma probe.

The use of a pinhole collimator means that the field of view, sensitivity, and spatial resolution of the system are highly dependent on the distance of the detector from the source. At distances from the source of less than around 5 cm, the camera's sensitivity and spatial resolution are comparable with those of a conventional gamma camera. The camera is easy to manipulate and can be positioned close to the patient. For example, in locating sentinel nodes in breast cancer patients, it is possible to position the camera close to nodes in the axilla.

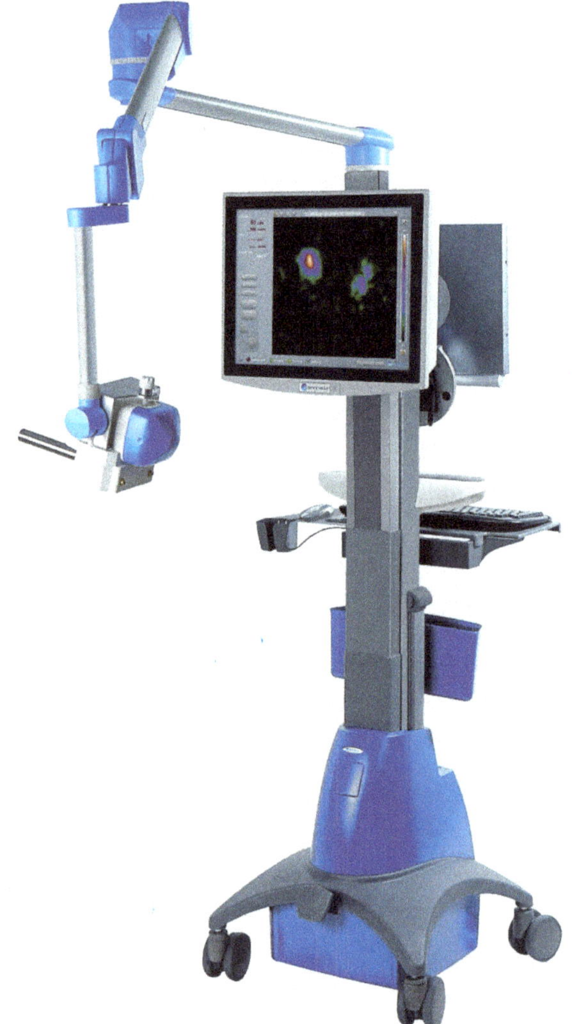

Figure 7.12 Sentinella 102 (reproduced with kind permission of Imaging Equipment Limited)

The camera is supplied with a QC jig and phantoms, and the Sentinella software includes tools for performing routine quality control tests and recording their results. The tests recommended by the manufacturer are given in Table 7.1.

QC testing is generally easy to perform, with clear instructions and the summary of results is well set out. However, the tests cannot be easily customised by the user. QC tools for the gamma probe are not included. These should be carried out

Table 7.1 Manufacturer recommended routine QC checks for the Sentinella 102 portable gamma camera

Test	Source and distance from collimator/camera	Frequency	Acceptance limit
System uniformity	Fillable phantom filled with ^{99m}Tc[a] at 30 mm	Weekly	<10% for both intrinsic and differential uniformity
System sensitivity	Small ^{99m}Tc source at 100 mm	Monthly	>27 cps/MBq (4 mm pinhole) >9 cps/MBq (2.5 mm pinhole)
System spatial resolution	Line source of ^{99m}Tc at 50 mm	Monthly	<16 mm (4 mm pinhole) <11 mm (2.5 mm pinhole)
Temporal resolution	Dead time and count rate at which response falls by 20% (R_{20}) are calculated using the two source method (IPEM, 2003a)	Every 6 months	Dead time <100 μs R_{20} >2210 cps
Intrinsic energy resolution	Small ^{99m}Tc source at 300 mm	Every 6 months	<17%
Pixel size	Phantom containing four point sources of ^{99m}Tc at 30 mm	Every 6 months	<0.2 mm

[a]The flood has no bubble trap and is difficult to fill. For frequent testing, it would be more convenient to have a small ^{57}Co sheet or to measure intrinsic uniformity.

according to guidelines published by the British Nuclear Medicine Society (Britten et al., 2005).

The frequency of testing and the stringency of the QC limits for a camera like this should depend on the use to which it is put. If the camera is being used for locating sentinel nodes during breast cancer surgery, the fact that its uniformity is not as good as expected from a conventional gamma camera does not mean that it should not be used. However, if the camera were to be used to produce diagnostic images, more stringent QC limits would be required.

7.7 Siemens IQ•SPECT

7.7.1 Introduction

The Siemens IQ•SPECT cardiac imaging system (Figure 7.13) utilises cardio-focal SMARTZOOM collimators with a cardio-centric orbit (Figure 7.14). A proprietary reconstruction algorithm is then used for image presentation. An advantage of IQ•SPECT is that it is used with a conventional dual-headed gamma camera with NaI(Tl) crystals (Siemens Symbia or Intevo). It is therefore more of a novel collimator and software combination than a novel gamma camera.

7.7.2 Operation

The two gamma camera detectors are positioned in a 76° configuration and rotate through 104° in a total of 34 projections. The front surface of the collimator is 28 cm from the isocentre. The patient lies supine on the imaging couch and the heart is positioned in the central magnification area of the collimator (the 'sweet spot'). Sensitivity in the 'sweet spot' is around four times greater than with low-energy, high

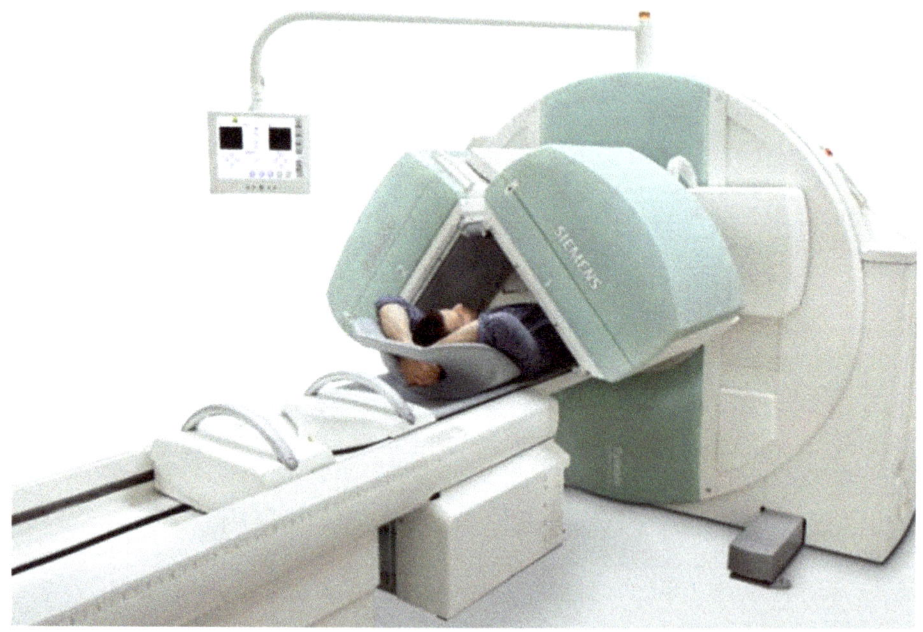

Figure 7.13 Siemens IQ•SPECT cardiac imaging system (reproduced with kind permission of Siemens Healthcare)

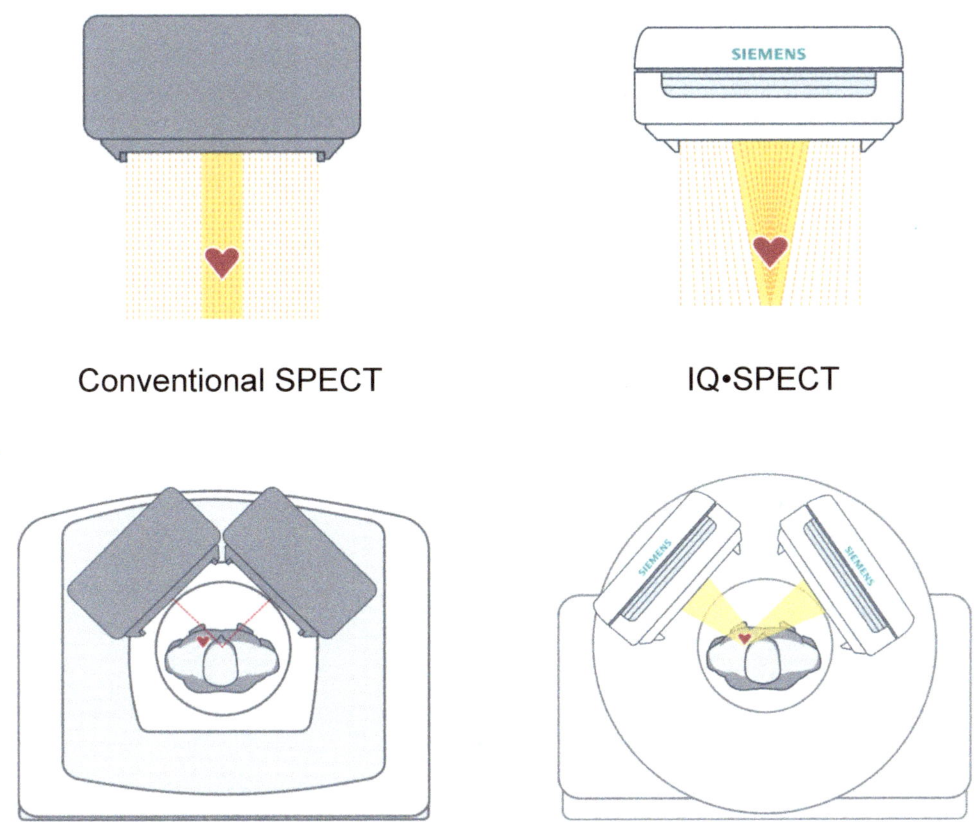

Figure 7.14 Cardio-focal SMARTZOOM collimator and cardiocentric orbit compared with conventional SPECT (reproduced with kind permission of Siemens Healthcare)

resolution (LEHR) collimators. This greater sensitivity allows for either a reduction in acquisition time, a reduction in administered activity or a combination of the two. It is important that the heart remains within the 'sweet spot' for the duration of the acquisition, as the sensitivity drops outside this area. Staying within the central magnification area will ensure that the noise characteristics in the dataset are representative of tracer distribution rather than differences in sensitivity across the detector.

Distortion corrected and uncorrected datasets are generated after acquisition. The distortion corrected dataset allows the user to review projections that appear analogous to conventional SPECT. This allows for review of patient motion and assessment of overlying bowel activity. Uncorrected data are used for image reconstruction with a proprietary iterative reconstruction algorithm. The algorithm models the geometric response and sensitivity of the collimator using a vector map of the individual IQ•SPECT system. There are inherent differences in the resultant images when compared with conventional SPECT, particularly with the non-attenuation corrected data. Slightly non-orthogonal photon trajectories are responsible, which results in photons travelling along different path lengths. This is in contrast to conventional SPECT where all collimator holes are perpendicular to the crystal. It is therefore recommended that a CT scan for attenuation correction be performed with IQ•SPECT studies.

7.7.3 QC tests

a) Routine QC

There are no QC tests specific to IQ•SPECT that are recommended by the manufacturer. The manufacturer states that, providing routine QC results are satisfactory using LEHR collimators, IQ•SPECT performance can be inferred to be satisfactory because SMARTZOOM collimators are a similar weight to the LEHR. The aforementioned routine QC procedures include:

- daily uniformity ('extrinsic sweep verification')
- centre of rotation
- multi-head detector alignment ('multi-head recognition').

These are in addition to assessment and calibration of intrinsic performance which is applicable to the performance of the system in general. As the cardio-centric orbit is dependent upon radial and tangential movements, the user may wish to devise their own tests in-house for routine IQ•SPECT QC, such as an IQ•SPECT scan of a point source positioned off-axis with the 'sweet spot' centred on the point source.

7.7.4 Acceptance testing

The procedures carried out for acceptance testing are mainly developed by users in-house. In addition to the procedures outlined in the routine QC section above, several other tests can be carried out to verify the performance of IQ•SPECT. These are outlined below.

a) 2D Sensitivity and uniformity

A two-dimensional image is acquired using a ^{99m}Tc or ^{57}Co flood source comprising 30–120 million counts. If a high count flood is acquired, the user may see a subtle image of a cross, representing the differences in sensitivity in different areas of the collimator. For this reason, a flood acquired using the SMARTZOOM collimator is expected to be slightly less uniform than for the LEHR collimator. In practice, absolute differences in non-uniformity of less than 1% between SMARTZOOM and LEHR collimators are observed.

b) 3D Sensitivity

A crude three-dimensional measure of sensitivity can be made quantitatively compared with an LEHR conventional SPECT scan. This is performed by undertaking IQ•SPECT and LEHR conventional SPECT scans of an anthropomorphic torso phantom cardiac insert. A fixed acquisition time is used and counts in the summed projections for the IQ•SPECT and conventional SPECT scans are compared. The counts in the IQ•SPECT study should be around four times that of the LEHR study.

c) 3D Uniformity

Given that the sensitivity of the detector varies across the field of view with the IQ•SPECT collimators, it is important that this variation in sensitivity is accounted for in the reconstruction. This can be tested using a Jaszczak uniformity phantom.

d) Anthropomorphic phantom

As is the case for some other dedicated cardiac imaging systems, myocardial perfusion images obtained with IQ•SPECT appear slightly different from those obtained with conventional SPECT. Therefore, it is advisable for the user to acquire phantom data using the anthropomorphic torso phantom with cardiac insert, showing both a normal distribution of activity in the myocardium and with defects introduced. This will allow the user to become accustomed to the differences seen in myocardial perfusion that are a feature of IQ•SPECT. As a result of the different photon path lengths travelled through tissue and thus differing attenuation between IQ•SPECT and conventional SPECT (as described in Section 7.7.1), it is important that the cardiac insert is imaged inside the filled torso.

e) Validation of gated data

In-house validation studies, presented at scientific meetings by some users, suggest that gated data such as left ventricular ejection fraction and cardiac volumes may not be comparable between IQ•SPECT and conventional SPECT. In order to investigate this, a dynamic cardiac phantom with variable ejection fraction is desirable. However, such phantoms are expensive and need to be suitable for purpose.

8 Testing of commercial nuclear medicine software

Philip Cosgriff

8.1 Introduction

Nuclear Medicine software can be broadly split into two categories: 'user written' software, also known as in-house-developed software, and commercial software. User written software in nuclear medicine has historically been produced for three main reasons. First, to overcome shortcomings and limitations of 'standard' image processing packages supplied by manufacturers. Second, to implement recently published clinical procedures and, third, in the context of local research, to analyse data in a completely new way. The first issue has been a significant problem in the past, but the situation has slowly improved to the point where the need for the user to write routine image processing software is now fairly rare. The second issue certainly remains, but manufacturers are now better at getting clinical software to market, so only those departments that need to implement newly published methods without delay are generally affected. The need to produce image processing software for research purposes will always remain unless, of course, the preparatory software development work is contracted out to a commercial company.

In IPEM Report 86, the focus was in-house developed software (IPEM, 2003a) and although what might be called 'traditional' in-house software development has continued to decline over the last 11 years, other forms have flourished. As this subject is much wider than nuclear medicine, IPEM's Informatics & Computing Special Interest Group (ICSIG) will address the software quality assurance and legal aspects in a separate publication. This chapter will focus only on the quality assurance requirements for commercial Nuclear Medicine software, providing specific practical examples that will supply the reader with the tools required to adequately characterise their system.

8.2 Definition and hardware configuration

Commercial nuclear medicine software may be defined as data acquisition and/or processing software produced by a commercial company for application within the specialty of nuclear medicine. Most commercial software is associated with the acquisition and processing of gamma camera images, but may also include many other non-imaging devices used in nuclear medicine (e.g. organ uptake probes, sample counters, whole body counters, radionuclide calibrators).

The main gamma camera vendors will supply a range of acquisition and processing software, usually on separate dedicated workstations connected via a local area network (LAN). Third party nuclear medicine software manufacturers generally only supply data processing software, since the stored (acquired) images can now be simply imported as raw DICOM files from the gamma camera system.

Third party workstations and servers are connected on the same network as those of the main gamma camera supplier and are also usually connected to the main hospital network, in which case, client software can be installed on any hospital computer (subject to hardware requirements), enabling nuclear medicine image processing to be performed at remote locations, including different hospitals within the same Trust. Security and other issues associated with connecting medical computer systems via local or wide area networks (LANs or WANs) are covered in IEC 80001-1 (IEC, 2010).

8.3 Testing of utility software

Utility software is a term used to describe small discrete applications (e.g. for calculating the gradient of a curve or the number of counts within a region of interest) that may be used on their own or be incorporated within larger clinical applications. Some basic testing of utility software is recommended as part of acceptance testing of a new computer system, or after a major upgrade of an existing one.

When a software upgrade is ordered or accepted, it is advisable to obtain a list of checks that will be performed by the supplier, and to request a copy of the test results summary sheet when the upgrade is completed. These tests will not normally include basic utility software, so it is prudent for the user to perform some rudimentary checks before putting the new system into routine clinical use. The issue of re-testing certain utility software following a 'major upgrade' of the nuclear medicine computer system has been recognised by the British Nuclear Medicine Society (BNMS) in its organisation standards, stating that certain programs (for data framing, image quantification, image arithmetic, time/activity curve arithmetic) should be specifically tested (BNMS, 2013). The testing of data acquisition software (static, dynamic, gated, single photon emission computed tomography (SPECT)) is covered in Chapters 3, 4 and 5 of this report and also a recent International Atomic Energy Agency (IAEA) publication (IAEA, 2009).

It is essential to have a test data set available so that the results of image processing can be compared to a reference. In the context of a software upgrade, such test data should be available from initial acceptance testing and it is useful to generate a combination of simulated and real data for this purpose (IAEA, 2009). The following simple tests use real data.

Counts within a static image

- Acquire a uniform static image using a pre-set count of 2 million and denote as image 1.
- Use the appropriate static image processing software to determine the total count in the image.
- Check that this is 2 million ($\pm 1\%$).
- If the system provides access to the DICOM (or Interfile) header, check that the total counts attribute for the image equals 2 million counts ($\pm 1\%$).

Counts within a region of interest (ROI)

- Define a ROI that covers the entire area of the uniform image (image 1) and denote as ROI 1.
- Define an ROI that covers approximately half of the area of ROI 1. Denote as ROI 2.
- Define an ROI that covers the remaining area of ROI 1. Denote as ROI 3.

Use a standard utility program to find the total counts within an ROI to calculate the values for ROIs 1, 2 and 3 when superimposed on image 1.

- Check that the counts within ROI 2 + counts within ROI 3 = counts within ROI 1.
- An error of $\pm 1\%$ would be acceptable, given the issues of defining accurate ROI manually.
- If the software has the facility to create a box-shaped ROI, this should be used.

Profile across a static image

Write a simple computer program to create a bar pattern image. For a 64×64 pixel image, write 16's into the first eight rows, 32's into the next eight rows, etc., producing a total of 8 bars. Use a standard profile generation program to create a one pixel wide profile across the image (i.e. perpendicular to the bars) and denote as profile 1.

- Check that the profile pattern is as shown in Figure 8.1 and that the *y*-axis (count) values are correct.
- Repeat for a 5 pixel-wide profile (profile 2) and check that the count values are five times that of profile 1.

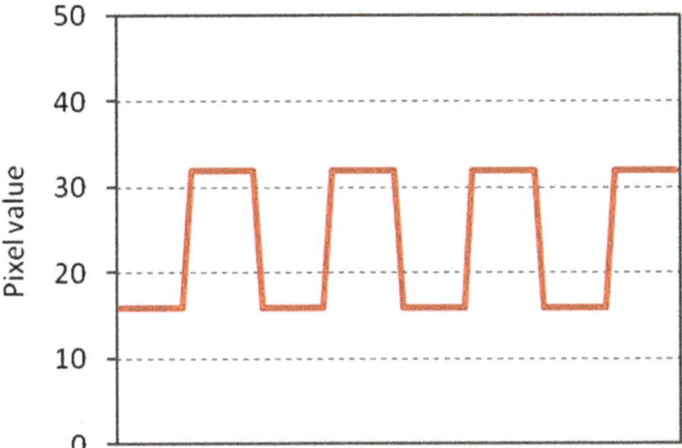

Figure 8.1 Profile 1 – the profile that should be generated by placing a 1 pixel wide ROI across the simulated bar pattern test image described in the text

If the suggested mathematic test object cannot be produced by local programming, a suitable physical phantom may be used in conjunction with a ^{99m}Tc or ^{57}Co flood source, but a standard commercial 4-quadrant bar phantom (for subjective resolution estimation) would probably be unsuitable as the number and spacing of the bars are suboptimal.

Time–activity curves from dynamic studies

Use simulated or real clinical data (i.e. a multi-frame dynamic study) for which the variation of counts with frame number within a particular ROI is either known (in the case of simulated data) or predictable (e.g. bladder ROI from a normal renogram study). Denote as Image Set 1.

To check the displayed count rate itself, write a suitable program to generate a 64-frame dynamic study (each frame having a 64×64 pixel matrix) in which the counts within each pixel are constant within a given frame, but which increase from 0 in frame 1 to 256 in frame 64, in steps of 4.

- Create a rectangular ROI covering the anatomical feature (in the case of real data). Denote as ROI 1.
- Create two ROI (ROI 2 and ROI 3) by dividing ROI 1 into two parts (not necessarily equal).
- Generate time–activity curves TAC 1, TAC 2, and TAC3 from Image Set 1, corresponding to ROI 1, ROI 2, and ROI 3, respectively.
- Check that the variation of count rate with frame number (time in the case of real dynamic data) is as expected. If you have used the above simulated data, the TAC will be a linear ramp function with an origin at (0,0) and an end point at (0,256).

- Add TAC 2 to TAC 3, to produce TAC 4, using the supplied standard curve arithmetic software.
- Check that TAC 4 = TAC 1. That is, display the two graphs on the same axes and check that they are exactly superimposed.
- Subtract TAC 2 from TAC 1, to produce TAC 5.
- Check that TAC 5 = TAC 3.

Miscellaneous

In addition to the above specific tests of acquisition and utility processing software, it is advisable to perform some other types of checks to confirm that an upgraded system is performing as expected (Section 8.5).

Results of individual checks should be signed/initialled and dated, and overall sign off ('OK for routine clinical use') should be authorised by the head of the nuclear medicine physics team. A summary of quantitative test results obtained should also be recorded electronically (e.g. in a spreadsheet). Clearly, when a customised or locally written program is itself updated, there are more stringent re-test requirements than outlined above.

8.4 Testing of clinical software

Commercial image processing software will generally be supplied with a disclaimer as part of the licence agreement. An example is as follows: "[Company name] takes no responsibility for any results and diagnoses derived from using the program described or from information in this manual. The users of the software are solely responsible for its use and for the resulting diagnoses".

If the software is Conformité Européenne (CE) marked, the manufacturer is claiming that the device meets all of the essential requirements of the relevant Medical Device Directive, including general ones that may represent add-ons to any 'blanket' disclaimer. Although the commercial manufacturer owes a duty of care in respect of patient safety, there is also an onus on the user to perform some basic benchmarking tests, especially for newly released software.

The most important testing of clinical software is validation, which is achieved by inputting standard (reference) test data for which the correct output is 'known'. Here, the 'correct output' is usually a quantitative physiological parameter (e.g. left ventricular ejection fraction [LVEF], relative renal function), but may relate to more subjective measures such as visible defects or other image features.

Suitable test data can be either simulated (i.e. physical or mathematical phantoms) or real clinical data, the latter being referred to as 'software phantoms' (Busemann Sokole, 1990). A database of normal and abnormal studies (approximately 10 of each) should be kept for each clinical program in routine use. It is also recommended to 'stress' any new or revised software using an additional set of 'extreme' data, which previously proved problematic using the old software. This usually relates to poor organ function (and associated high background counts), but could also be due to unusual anatomy, etc.

Results from the new software can thus be directly compared with the old (assuming that the new program will read the old data) and an investigation made if the results are more than 5% different (IAEA, 2009). However, it should not be automatically assumed that the new software is faulty if the result produced is more than 5% different, as it is possible that a new algorithm may produce a result that is closer to the true answer. For some physiological parameters obtained from nuclear medicine imaging, the true answer may not be known with any degree of accuracy (i.e. the 'gold standard' test may not have been performed). Intra- and inter-observer variability should also be systematically tested. Intra-observer variability should be less than 3% and inter-observer variability should be less than 5% (IAEA, 2009).

Apart from use of locally obtained data, it is also advisable to include at least five of the datasets available from IPEM for the most recent relevant software audit, as the median result from many participating centres is likely to be closer to the true result than that obtained by any individual centre.

Reference databases are also available from other sources. For example, a database of over 100 clinically documented (but not validated) adult mercaptoacetyltriglycine (MAG3) renography datasets covering a wide range of renal function has been assembled by a group working under the auspices of the International Scientific Committee on Radionuclides in Nephro-Urology (ISCORN). Files can be downloaded in DICOM or Interfile format (ISCORN, 2011). It should be noted that use of these data for testing renography software is subject to a licence agreement.

The same testing schemes applied to commercial clinical software would of course also apply to clinical software developed in-house.

8.5 Management of software upgrades

8.5.1 Clinical software

The following refers mainly to commercially supplied software, but the clinical validation aspects would apply equally well to clinical software developed in-house. For the latter, testing of different software components and modules would also be specifically tested as part of the development process.

Upgrades to commercially supplied software are generally offered as part of a comprehensive maintenance contract. Some are offered at no additional cost while others are not. Upgrades to the main user interface (i.e. the database used to select patient files for processing) are always included in the contract as the manufacturer has a vested interest (from an ease of maintenance point of view) of having all users on the same version of the 'patient viewer'.

It is important to properly evaluate new clinical software to decide whether to implement it, as the decision is not automatic. Some upgrades may improve aspects of the software that are not even used locally (e.g. factor analysis in a renography program) and it may not be worthwhile going through the various local change management procedures for very little, if any, benefit. The main steps in the process are thus as follows:

1. Critically evaluate what is being offered. This would typically involve a demonstration of the new software by the supplier. It should also include reading the Release Notes for the program in question in order to identify exactly what has changed between the existing version and the new one, which will typically include bug fixes and other general improvements. You may have to request this document specifically, as it probably will not be included with the new user instructions.
2. Check that the supplied user documentation is of the required standard.
3. Test using real clinical data, either stored locally or obtained from professional bodies (Section 8.4).
4. Test for inter-observer variability (Section 8.4).
5. Test against existing program, if one exists.
6. Train the users, update their training records and update the user instructions when the software is formally introduced into clinical service.
7. Keep a copy of the old program in case of unforeseen problems. This will have a different version number to the new one referred to in the departmental clinical procedure manual, but should be hidden from the user to avoid inadvertent use. This is a job for the System Administrator, who will have 'admin rights' when logged in and be able to customise the view of the system (including list of available clinical software) that individual users have.

In some cases, there may be an expected systematic difference between the output of the old and the new software. For example, there may now be evidence that the old method tended to underestimate the parameter in question by 5–10%.

8.5.2 Operating system upgrades

When an upgrade of the operating system (or main user interface software) is planned, it is advisable to obtain a list of checks that will be performed by the supplier, and to ask for a copy of the test results' summary sheet when the upgrade is completed.

It is not usually necessary to test individual manufacturer-supplied image analysis programs (e.g. renal, cardiac, brain) following an operating system (or user interface) upgrade, as this compatibility issue should clearly be fully tested by the manufacturer. However, it will be necessary to update individual clinical procedures (image processing aspect) if they refer to a version of the main patient image viewer ('front-end') that has been updated. What tends to be forgotten (by the manufacturer) are customised scripts for dealing with specific local situations (e.g. special printers, back-up devices) and local software written using a scripting or high level language, so it is obviously important to re-test these within the new environment.

For free 'routine' software upgrades initiated by the supplier, especially operating system upgrades, it is advisable to gain sight of the service engineer's job sheet before authorising (what is the upgrade designed to do?), and check that the upgrade has been successfully carried out at other UK sites. In any event, it is prudent to ensure that a member of the department's physics team is in attendance during the engineers' visit. Most importantly, all patient data (not already archived) and any customised versions of clinical software should be backed up before the upgrade starts.

Main user interface or operating system upgrades can have unexpected consequences, so it is advisable to perform some basic tests of overall system performance to make sure that everything is still working as it was before. A recommended checklist is given in Section 9.6.4.

Compliance with recommendations in this sub-section should satisfy the BNMS organisation audit requirements on the subject of required procedures and documentation following 'computer system upgrades' and 'major software revisions' (BNMS, 2013).

9 Life cycle of a gamma camera

Mike Avison, Philip Cosgriff, Ewan Eadie, Glen Gardner, Joanne Kerry and Richard Lawson

9.1 Introduction

The final chapter in this report widens its scope to consider topics that may directly or indirectly impact on gamma camera quality control. Here we discuss the gamma camera's life cycle, starting with procurement considerations and concluding with equipment replacement. The detail contained in this chapter comes from the combined experience of the contributing authors and, where possible, the reader is directed to additional published material.

9.2 Planning

9.2.1 Equipment specification

The first step in procuring a gamma camera is to take great care with equipment specification. In many organisations, once the specification is complete, there will be little opportunity to influence which system is purchased. The reason for this is that the tendering process must be shown to be fair to all companies when public money is being spent. Equipment specification is an important part of the overall tendering process (Section 9.3) and it is useful to divide the generation of a tender into separate phases.

a) Discovery of purpose

It may seem obvious that a replacement gamma camera will be required to perform the same functions as the previous one. However, the new machine will affect a department's ability to meet service needs for the next 5–7 years (the planned life of the camera), so it is essential to consider what new types of scan may be performed in the future. This will usually involve arranging meetings with existing and potential service users to gauge the demand for new services. If the purchase is for a new department or to fulfil a specific new demand (e.g. cardiac imaging), it will be more obvious that careful consideration is required to purchase the most suitable camera.

In addition, an examination of the stresses placed on the current service should be made and thought given to how different design features might ameliorate the current issues. For instance, if performing a lot of one type of scan, myocardial perfusion for instance, and current demand is exceeding capacity, some of the different technologies available could significantly improve departmental throughput (Chapter 7). Other productivity improvements might be gained by, for instance, having access to the processing computer from multiple PC workstations around the department or hospital at large. Another feature that suits some users is the ability to stack up processed results in separate workflows for specific reporters to access. It is also worth considering whether a change to existing scanning protocols might be appropriate with the potential availability of additional features on the camera (e.g. a pinhole collimator or X-ray attenuation correction).

b) Making a list of important features

Before starting the formal tendering process, it is advisable to look carefully at both standard and special features available on each of the gamma cameras under consideration. Items such as collimator choice can often be overlooked but have a fundamental impact on image acquisition. Informal discussions with company representatives, visits to other centres, and conference trade exhibitions are good places to gain information. It may also be helpful to pose some questions on a forum such as the medical physics and engineering mailing list (www.jiscmail.ac.uk). The manufacturer will usually be happy to provide a list of all centres that have purchased a particular option, if direct contact would be preferred. It is useful to ask current owners of the cameras being considered if there are any problems with specific features. Gaining this kind of information is useful in developing the specification because it might lead to phrasing part of the specification in a particular way. For example, a feature may be aimed at making scanning more time-efficient but using it may lock out some of the options required in practice. The gain in efficiency may not be worth the loss in flexibility. This kind of research may also inform the list of things to be demonstrated at formal site visits. One example of such a situation is contoured single photon emission computed tomography (SPECT) acquisition. Most manufacturers offer this, but not all have implemented it in such a way that both heads contour independently when the detector heads are in 90° configuration. Another current issue is resolution recovery SPECT reconstruction. Most manufacturers offer this either in the standard package or as an option, but there is little objective evidence on the effectiveness of individual software products. Investigating this level of detail can avoid disappointment later. All manufacturers offer a range of collimators for different purposes but design criteria differ, so it is worth comparing specifications for sensitivity and resolution of available low energy collimators (for ^{99m}Tc) as well as medium and high energy collimators if required. Radionuclides such as ^{123}I and ^{81m}Kr are known to cause specific design issues and so, if there is a requirement to use these, then particular care should be taken to ensure that suitable collimators are included.

c) Drawing up the specification

If the open tendering (Section 9.3.1) process is selected, no criteria can be used in assessing the tenders which were not mentioned in the specification. When it comes to quantitative aspects of the specification, most, if not all, manufacturers quote performance in terms of the test procedures laid down by the National Electrical Manufacturers Association (NEMA, 2012). It is important to be aware that these tests are not performed independently, and not even with external scrutiny, so there may be some inaccuracy in the published results. The British Nuclear Medicine Society (BNMS) have produced a useful questionnaire for gamma camera purchase which is available on their web site (www.bnms.org.uk). All manufacturers should have available their response to the standard questions for each of their camera models, so it is useful to obtain a copy of their response at this earlier stage too. Reading a copy of the basic questionnaire before drawing up the specification will also help avoid important omissions.

The specification should include both quantitative and qualitative criteria – more general statements focused on the clinical need to be fulfilled. This can be useful at the assessment stage in case the quantitative results are in some way misleading.

Some of the items on the specification are given essential status, in which case any tender that fails to meet one of these, will be excluded from the process. Other items on the specification require a quantity in the reply (e.g. sensitivity or resolution) in which case these items may be used in the assessment scheme to score the cameras.

If asking for remote access to the processing computer (often termed "floating licences"), ensure that it is clear which functions are required for each concurrent user because some features may not be available under floating licences. Manufacturers vary considerably in the way they cost this. Some of the charging schemes are listed below:

- Separate charge for each module of standard software.
- All standard software included but all options charged for.
- All manufacturers own software included, only third party supplied options charged for.

The difference between these models, where three or four floating licences are required, can amount to tens of thousands of pounds.

As well as the performance specification, consideration must be given to the environmental requirements of the system, such as room size, floor strength, power requirement, etc (Section 9.4.1). This area should not be underestimated and the process can take much longer than expected, particularly if a private finance initiative (PFI) is involved. If a hospital is part of a PFI, then the buildings will be leased from an operating company, so agreement will be required for any alterations to the camera room or its services and the implications that this has for future maintenance of the facilities. It is essential to bring together all interested parties as soon as possible. This may include representatives of the user department, purchasing department, finance department, radiation protection, medical physics support, and of course the facilities department. Once a tender has been accepted, a representative of the successful camera supplier should also be included.

Once the specification has been drawn up, it should be used to make an assessment scheme, which is most easily done using a suitable spreadsheet. All of the important items should be listed in the specification and space left to allocate points to indicate how closely the requirement is met by each potential supplier. It may also be necessary to use weighting factors to reflect the relative importance of the respective items. The weighted sum of the points for all items can then be used as an overall score to compare each system.

9.2.2 Site visits

Whichever method is used for tendering, the purchaser will usually wish to make visits to existing users of the equipment being considered, which is usually organised by the respective suppliers. It is clearly useful to choose a site that conducts a similar range of procedures and where the equipment includes at least some of the specific options being considered.

Purchasers may also wish to make their own arrangements to visit a site "unofficially" (i.e. without the presence or assistance of the manufacturer). It is difficult to assess equipment reliability during an official visit, as the manufacturer will try to steer clear of sites where there have been recent problems, and local staff may also feel discouraged from giving a negative assessment in front of the manufacturer's representative.

Before the first site visit, it is essential to draw up a checklist of priorities. Make sure the manufacturer's representative knows what is on the list, especially the high priority items.

It is useful to take a multidisciplinary team to the visit, such as a reporting radiologist/physician, a physicist, and a technologist. Hosting sites will often arrange for each staff group to pair up with a person from their own discipline for at least a part of the visit.

If high patient throughput is a priority, it will be helpful to time some operations such as collimator change and SPECT-CT set-up. Ease of collimator change can make a big difference between systems and this can often only be discovered during a site visit. In the authors' experience, the manufacturers usually quote SPECT-CT scan times by simply adding the emission and transmission scan durations, but some systems take a considerable amount of time to reconfigure the heads and/or move the bed/patient between scan positions.

Similarly, systems that appear very flexible for seated or standing patients may take a long time to perform the required detector head reconfiguration, so the claimed flexibility is of limited practical value.

It is rarely possible to scan a phantom at a site visit but a foam cross-section of a patient can be useful to observe how well the camera performs autocontouring during SPECT acquisitions.

9.3 The procurement process

For NHS organisations (in the UK), there are two routes for purchasing:

9.3.1 Purchasing by open tender

The usual method is to seek competitive tenders by advertising in the *Official Journal of the European Union* (OJEU). A "Summary of Need" is required to send to OJEU, the technical part of which is very simple (a few lines describing what kind of equipment is required and what kind of clinical work is to be performed). The hospital's local finance/purchasing department will want to provide the structure and they will have some standard phraseology to accompany the technical part.

Prospective suppliers will then respond to the OJEU advertisement and request the tender documents that represent the detailed description of exactly what is required. The detailed specification is only a part of the tender documentation; other parts will cover financial arrangements and civil works details.

The companies who wish to tender will usually be sent a pre-qualification questionnaire, which seeks to eliminate suppliers who are clearly unsuitable.

The remaining companies are then asked to send in their tenders. Once the tenders have been received, the manufacturer's specification is compared against the user-written specification (probably based on the BNMS tender questionnaire) using objective criteria. The assessment document must clearly directly relate to the user specification. A purchaser cannot assess equipment on criteria that were not mentioned in the tender documents. Although suppliers do not normally get to see the assessment scheme, they may challenge the process if deemed to be unfair and in such cases, the company's lawyers will require access to the assessment scoring scheme.

There are minimum time scales that must be adhered to for each part of this process. The hospital's finance department will advise on these but usually the advertisement must be in the OJEU for a minimum of 1 month before the opportunity to express an interest closes. When the shortlisted companies have qualified, there is then usually a further period of 4 to 6 weeks for them to complete and send their tender bid. No company can be given the opportunity to extend a deadline. If a company accidentally fails to respond in time and the purchaser wants that company included, the situation becomes problematic, and the hospital's finance department will explain the options available.

9.3.2 Purchasing from supply chain

NHS Supply Chain (NHSSC) is a division of the NHS in the UK (http://www.supplychain.nhs.uk/) whose stated aim it is to ensure the NHS gets best value for money when purchasing goods and services (by using the collective bargaining power of a huge organisation), and takes some administrative strain off hospital trusts in the process.

At regular intervals (once per year for most large capital items), the NHSSC goes through an open tender process for each type of equipment. Therefore, a hospital trust planning to purchase a piece of equipment can find out the latest prices and place an order through supply chain, which is much simpler and quicker than using the site-specific open tender option.

However, opinions vary as to which route is the best option. The prices available through NHSSC are generally 'base level systems' without optional extras. Any extras are therefore not being purchased through competitive tender with this route, and the overall system cost could be higher. However, it is also possible for a company to come up with a favourable bundled price, including any extras that the user wants, so it is well worth asking each possible supplier for their price for an overall package rather than just their base system. In some circumstances, this can allow the user the flexibility to obtain a package that meets all of their requirements at a better price than would have been possible through the rigid open tender process.

Even though a user specification is not needed for this option, it is still a good exercise to go through to ensure that proper consideration has been given to all of the technical and logistical aspects as well as the all-important issue of ongoing maintenance costs.

Checklist	Description
Summary of need (only if going to OJEU)	Brief statement about equipment required
Specification	More detailed list of attributes
BNMS questionnaire	Ask the manufacturers for this as part of the tender
Site visit checklist	Scan set-ups, etc. that you wish to see at all visited sites
Assessment scheme	Objective means of comparing tenders against the specifications submitted

9.3.3 Leasing the equipment

Leasing is an increasingly attractive option for Trust's having difficulty allocating capital funds to purchase expensive medical equipment outright. This can be done through a leasing company or by direct negotiation with the equipment supplier, but is now most commonly organised as part of a comprehensive managed equipment service (MES). An MES arrangement is only a viable option if there are economies of scale, so numerous items of equipment would need to be included. In a large Trust, it may be financially viable to include just one modality (e.g. ultrasound) but would more typically include all 'radiological' equipment (CT, MR, US, NM) to achieve the largest overall long-term cost saving. The closeness of the relationship between the local users and the chosen MES provider is obviously the key factor in obtaining equipment that meets the changing service requirement.

Large multi-national medical imaging companies such as GE Healthcare, Siemens and Philips have their own MES programmes, and claim to offer multi-vendor procurement, but there are also dedicated third party MES companies that are completely independent of the equipment manufacturers.

9.4 Equipment installation

9.4.1 Room requirements

As part of the tendering process, it is essential to require each potential supplier to provide a suggested room layout in order to ensure that their gamma camera will actually fit in the available room space. However, after the successful system is chosen, there should be an opportunity to optimise the room layout taking account of items such as:

- access for patients who may be in a wheelchair or a hospital bed,
- accessibility to services such as medical gases and the washbasin,
- space required for collimator storage and changing,
- view of the patient from the position of the operator's console,
- position of other equipment such as an uninterruptable power supply,
- does the camera position enable a point source to be positioned sufficiently far away for intrinsic measurements (5.5 times the diagonal of the field of view (FOV))?

This optimisation requires discussion between the staff who will use the equipment and the supplier's representative, the process being much easier if staff have previously had an opportunity to visit another site to see the equipment in routine clinical use. When the room layout has been optimised, the supplier will be able to provide a detailed technical drawing showing the installation requirements for their equipment.

Suppliers will sometimes be able to provide a 'turnkey' solution where they provide a complete installation service. This may include everything from removal of any previous equipment, necessary building, electrical and plumbing works, re-decoration and installation of the new camera. Alternatively, the hospital may prefer to have the camera supplier responsible only for installation of their equipment and contract elsewhere for the rest of the work. In either case, it must be made clear who is responsible for which part of the works and whose budget the money will come from. Care must be taken to consider all aspects of the camera installation from the hospital's point of view, not just those listed in the manufacturer's installation requirements. The following is a list of items to consider, although some may not be required in all installations.

a) Removal of any existing equipment

If the old camera is leased, it will have to be returned to the lease company, but if not, the hospital will have to dispose of it. If the old camera has some second-hand value, then there are companies that specialise in used radiology equipment who may be able to take it away, refurbish it, and find it a new home. If it has to be scrapped, there will be costs associated with disposing of the equipment appropriately. In any case, there will be a cost for disconnecting the equipment, removing it and transporting it away from the site, so it needs to be clear who will cover those costs.

b) Building works

If the new camera will be replacing an existing one of similar size, then very little building work may be required. However, there should always be an allowance for 'making good' the floors and walls after removal of the old equipment. If the existing room layout does not suit the new equipment, it may be necessary to move walls or doors or even to extend the department. This obviously makes the installation more complicated and will have an impact on surrounding areas, so it will need more planning.

c) Doors and walls

Radiation shielding provided by existing or new walls must be taken into account. If the new camera includes a built-in CT system, then appropriate lead shielding must be included in walls and doors. However, even when CT is not included, it is worthwhile (but not essential) to include about 2 mm of lead in the walls and doors, otherwise sources (such as patients) in other rooms can interfere with the camera operation when the collimators are removed. Doors need to be wide enough to allow easy patient access in a wheelchair and also for the largest (e.g. intensive therapy unit (ITU)) hospital bed. The doors need to be lockable as it is sometimes necessary to leave radioactive sources unattended in the room (e.g. for long duration QC tests).

d) Lead-glass screen

If the camera includes a built-in CT scanner, then the operator will need to retire to a suitably shielded area during the CT exposure. With some low dose CT units, it is sufficient to have a lead-glass screen in the gamma camera room with the acquisition computer positioned behind this, so that the operator can sit behind the screen during the X-ray exposure of the patient. However, for higher dose CT systems, it will be necessary to have a separate control room with a door and a lead-glass window. In this case, an intercom to communicate with the patient during the scan will be required and, in some cases, a closed circuit TV may also be needed. The supplier will advise whether a separate control room is needed for their system.

e) Floor construction

The floor loading of some gamma cameras can be considerable (up to about 3000 kg) and so floor construction must be able to take the weight. This is not normally an issue for departments on the ground floor but the hospital estates department will be able to advise. If floor strengthening is required, it can be a considerable extra expense for departments, especially for SPECT/CT cameras. In addition, the floor needs to be flat and level and some manufacturers require it to be level to within 1 mm per metre in certain critical areas.

f) Wall and floor finishes

Since it is likely that the camera room will be used for administering radiopharmaceuticals to patients during some imaging studies, the possibility of spills of radioactive solutions must be considered. It may also be necessary to disinfect the room from time to time if infectious patients have been imaged. Therefore wall surfaces must be wipeable, non-absorbent, and suitable for cleaning with hypochlorite solution. Floors must be covered with an impervious material such as vinyl with welded joints and coved up the walls. The material should conform to appropriate standards for radioactive decontamination.

g) Ceiling

There are not usually any specific requirements for the ceiling, but it may be worth considering installing a backlit image on the ceiling above the patient imaging table to give patients something to look at. Consideration should also be given to what is above and below the gamma camera room, and the radiation dose that may be obtained in those areas, especially if a SPECT-CT system is being installed.

h) Services

The room may require medical gases such as oxygen, anaesthetic gas, compressed air, and suction depending on the clinical need. One or more emergency call buttons may be required in easily accessible locations.

i) Heating and ventilation

Gamma cameras require a stable temperature in order to avoid expensive damage to the crystal. Manufacturers generally recommend that the temperature should remain between 20℃ and 25℃ and that the rate of change should not exceed 3℃ per hour. Since gamma cameras can produce considerable heat output (up to 3 kW for some models) continuously day and night, it is important that this requirement is highlighted early on in the design stage. Humidity should be 40% to 60% non-condensing and should not change by more than 5% per hour. Forced air extraction may be required when some radioactive gases are used.

j) Electrical supply for the camera

The manufacturer will specify the electrical power required by their gamma camera and whether it needs a three-phase supply or a single phase supply. There should be an identified earth reference terminal, an incoming mains isolator switch, and one or more emergency stop buttons. The gamma camera and the associated acquisition workstation should be supplied through an uninterruptible power supply (UPS). The purpose of the UPS is not to keep the camera and its workstation going for long periods, but simply to bridge the gap of a few seconds whilst the hospital's emergency generator starts up following a failure of the incoming mains supply. In hospitals where the emergency back-up cannot be completely relied upon (especially in situations where mains power is frequently interrupted), it is advisable to use the time provided by the UPS (usually at least 15 min) to simply shut down the cameras and associated computers in an orderly fashion and only to do a restart when the situation has completely stabilised.

If there is no UPS, then even a brief interruption to the mains is likely to cause the gamma camera and computer to turn off and therefore any study that is currently being acquired will be lost. It is important that the UPS should automatically switch back to the mains supply as soon as this is restored, otherwise if there is a mains failure during the weekend, the battery in the UPS can run down and the gamma camera may have been off for several hours when staff return on the Monday. This is undesirable because after turning on again, the camera can require a long stabilisation period before it is ready for use.

k) Other electrical power outlets

In addition to the supply for the gamma camera, there must be sufficient socket outlets located around the room in convenient locations to run things such as service equipment, additional computers, patient monitors, and other portable equipment. Fused spurs will also be required for any fixed equipment such as a ceiling mounted patient hoist.

l) Lighting

Good intensity lighting with colour correction is required to facilitate patient injection. However, it is also necessary to have a dimmer facility to dim lights to a low level when required during scanning. If a backlit panel is to be installed in the ceiling, do not forget that this will also need an electrical supply and a suitable switch.

m) Networking

Computer network points (either wired or wireless) will be required for the gamma camera and acquisition computer as well as any other computers anticipated now or in the near future. The manufacturer may require a dial-in access to the network from outside for their own service and system diagnostic purposes and this will need agreeing with the hospital IT department.

n) Trunking

Suppliers will specify what trunking is required to carry interconnecting cables for their system. This is usually wall mounted but some systems also require additional floor trunking to take cables to the camera gantry or the patient couch. In this case, provision must be made for channelling the trunking into the floor before the vinyl is laid. The trunking must have a metal lid strong enough to take the weight of people and equipment moving over it. In planning the room layout, it is a good idea to avoid having to wheel heavy collimator carts over the position of any floor trunking since carts can weigh up to 500 kg when loaded. Before the camera is installed, the floor trunking will have to be fitted and the floor covered with vinyl everywhere apart from over the trunking. Then the installation engineers can easily remove the lids and place their cables in the trunking. When the installation is finished, the trunking lids can be screwed down and covered with vinyl which must then be welded to the surrounding vinyl to make an impervious join.

o) Plumbing

A basin for hand washing will be required. Consider fitting this with hands-free taps operated by a proximity sensor.

p) Furniture and fittings

The gamma camera room, and control room if there is one, will require additional furniture such as shelves, wall cupboards, benches, under-bench cupboards and drawer units, computer desks, and chairs. Also, do not forget important small items such as soap dispenser, paper towel dispenser, glove dispenser, plastic apron dispenser, pedal bin, coat hooks, white board, clock, etc.

q) Radiation protection

The gamma camera room may need to be classified as a supervised or controlled area because of the use of radioactive materials and so suitable fixed signage at the entrance will be required. If the camera also has a built-in CT system, then the room will also become controlled due to the use of X-rays. This will require a suitable illuminated X-ray warning sign at the entrance.

r) Access

Some thought must be given to how contractors will gain access to the site during the installation without disrupting work in the rest of the department too much. In addition, a suitable access route for the new camera to be delivered must be agreed. Suppliers will advise on the requirements for their own cameras, but narrow doorways, tight corners or small lifts can cause problems.

9.4.2 Scheduling the installation

Scheduling of each stage of the installation process is clearly important because of the impact that it will have on operation of the rest of the department, and possibly the rest of the hospital. If the gamma camera that is to be replaced is the only one in the department, then there will be a period when no imaging can be carried out and alternative arrangements may have to be made. If the department has more than one camera, then it should be able to continue working during the replacement of a camera, but the workload will have to be adjusted to take account of reduced equipment availability.

The supplier will usually want to confirm a date for delivery of the new camera well in advance so that they can book their installation engineers for the required dates. Therefore, any slippage in the date that the room is ready may mean that the engineers are not available due to other commitments. One advantage of a turnkey solution here is that everything is under control of the one supplier who will appoint a project manager. Once the room is finished, installation and adjustment of the camera will usually take a few days, after which acceptance testing can begin.

9.5 Acceptance testing

The acceptance of a new gamma camera is a crucial part of the procurement process and adequate time should be earmarked, before commencement in clinical use, to ensure a comprehensive programme of testing is undertaken. As a guide, 1 week should be set aside for a multi-detector SPECT system, with additional time required for systems with a CT component.

There are three objectives when planning the acceptance tests of a new camera. First, verifying that the camera performs as per the manufacturer's specifications. This will involve reproducing the NEMA tests used by the manufacturer to ensure that the system meets the manufacturer's specifications. Second, setting the baseline for all future QC tests. These are unlikely to be the same as the NEMA test protocols and more likely to be less complex and use the pre-existing equipment within a department. Lastly, tests to verify the functionality of the camera are performed to ensure that all of the safety features are functioning and that the system performs as expected within the clinical setting.

9.5.1 Verifying manufacturers' specifications

All manufacturers will produce a specification document detailing how their machine performs under the testing conditions as described in the NEMA protocols. This will include tests of:

- intrinsic resolution,
- intrinsic spatial linearity,
- intrinsic uniformity,
- intrinsic energy resolution,
- intrinsic count-rate performance.

These tests are aimed at demonstrating the ultimate capabilities of the detector and will necessitate the use of special phantoms to ensure the results are comparable to the factory specifications.

During installation, it is usual for the manufacturer to perform a selection of these tests, normally intrinsic resolution, intrinsic spatial linearity, and intrinsic uniformity. It may therefore be possible to work alongside the engineer and accept the results of those tests. Some NEMA tests are however impractical to incorporate into an acceptance test schedule, for example, the intrinsic count-rate test requires a specific source geometry which limits the test to one detector at a time, resulting in an acquisition time of 2 days per detector.

9.5.2 Setting baseline for routine QC testing

Whilst it is important to ensure that the system meets the specifications, the end user may be more interested in how the system performs in a more clinical setting. For example, a system may have a claimed intrinsic 20% count-rate loss of 75 kcps, yet with the introduction of a collimator and scattering material, the 20% loss rate may well fall to 20 kcps. It is therefore important to also examine the system performance within a realistic setting.

There are several publications giving advice on which tests should be included in an acceptance test schedule, including the IAEA and EANM. Table 9.1 provides details of the acquisition protocols described in previous chapters of this report (Section 9.6.1) and indicates which tests are NEMA requirements. As mentioned previously, NEMA tests are very specific and require specific equipment. The test methods described in this report cannot be substituted for the NEMA test methods. However, it is necessary to carry out the additional tests described here using the same methods which will be used for daily, weekly, monthly or annual QC to obtain baseline values for these test parameters. It should be noted that if a system fails a test, then the manufacturer should make repairs before any further testing is carried out, as, depending on the nature of the fault, the entire process may have to be repeated.

9.5.3 Functionality

In addition to testing the imaging quality of the detectors, the user will also have to ensure that the functionality of the system is as expected. These tests will include verifying the operation of:

- all emergency stops (defining which buttons stop motion and which power off system and/or CT)
- emergency bed extraction (can the bed also be lowered)
- patient collision system (check every collimator at several gantry angles)
- auto-contouring system (is there a "look ahead" function to avoid feet for example)
- collimator exchange (is the loading system robust enough for daily use, multiple attempts at loading each collimator set)
- gantry motion (move gantry through all geometries and maximum travel, listening for loose particles or unusual sounds, can the bed collide with the detectors)
- detector shielding (how susceptible is the detector to radiation sources outside the field of view)
- collimator integrity (collimator hole angularity test to check each collimator for poor construction or damage)
- uninterruptible power supply (where fitted, does the UPS maintain power to the detectors).

It should be noted that, if the system requires a major component replacement or the system is relocated, a further set (or appropriate subset) of acceptance tests will be required to set a new baseline for future quality controls.

9.6 Maintaining the equipment

9.6.1 Frequency of quality control tests

Table 9.1 provides a guideline to the quality control tests mentioned in this report and their recommended frequency. More detailed information can be found in each of the referenced sections. The tests, and their frequency, recommended here are not absolute and the reader may find that some of these tests need to be performed more or less regularly. This will depend largely on the clinical procedures carried out on the gamma camera, and the medical physics expert (MPE) should develop an appropriate quality control strategy. This report can be used as a reference to aid the MPE in the decision process.

Table 9.1 Summary of quality control frequency with cross-referencing

Measurement	Quantitative / Qualitative	Intrinsic / System	Cross-reference (Section number)
Daily			
Uniformity	Qualitative	Either	3.3
Weekly			
Uniformity	Quantitative	Either	3.3
CT: Image noise	Quantitative	n/a	6.1.2
CT: Number accuracy	Quantitative	n/a	6.1.3
Monthly			
Sensitivity	Quantitative	System	3.7
Whole-body exposure time correction	Quantitative	System	4.4
Centre of rotation	Quantitative	System	5.5
Consistency of angular response (optional)	Quantitative	System	5.7
Quarterly			
Whole-body count rate variation	Quantitative	System	4.3
Six-monthly			
SPECT-CT registration	Both	System	6.2

Table 9.1 Continued

Measurement	**Quantitative / Qualitative**	**Intrinsic / System**	**Cross-reference (Section number)**
Annually			
Resolution	Quantitative	Both	3.4
Energy resolution	Quantitative	Intrinsic	3.8
Multiple window spatial registration	Quantitative	Intrinsic	3.9
Reconstructed SPECT uniformity	Qualitative	System	5.9
Total SPECT performance	Qualitative	System	5.11
CT: Radiation dose	Quantitative	n/a	6.1
CT: In-depth image quality	Qualitative	n/a	6.1
Resolution (optional)	Qualitative	Both	3.4.1
Acceptance \ Baseline*			
Uniformity	Quantitative	Both	3.3
Spatial linearity	Quantitative	Intrinsic	3.5
Count-rate performance	Quantitative	Intrinsic	3.6
Whole-body system non-uniformity	Quantitative	System	4.2.1
Whole-body resolution	Quantitative	System	4.2.2
Collimator hole angulation check	Qualitative	System	5.6
Reconstructed SPECT resolution	Quantitative	System	5.10
Effectiveness of CT attenuation correction	Quantitative	System	6.2.4

*These tests should be carried out at acceptance testing in addition to all of the other tests mentioned in this table.

In addition, this summary should be used to compliment the quality control and safety tests recommended by the gamma camera manufacturer and is not designed as a replacement. Indeed, the gamma camera manufacturer's procedures should be carried out as instructed to help ensure safe operation. Although not mentioned in Table 9.1, the MPE should also consider which quality control tests will be required following any changes or repairs to the gamma camera or its associated software (Section 9.6.4).

9.6.2 Action thresholds

An important feature of any quality assurance programme is that it is a closed loop process, which means that if any quality control measure indicates that there is a problem, then remedial action must be taken to rectify this and bring it back within control. There is no point simply recording daily, weekly or monthly measurements of parameters such as uniformity or SPECT centre of rotation unless each measurement is examined and appropriate action taken. Possible actions might be one of the following:

1. No remedial action required. The camera is fit for normal use. This will be the case if the measurement is within expected limits.
2. Take no immediate action, but assess the situation again after the next planned measurement (e.g. next day). This will typically apply if the result in question is only just outside the expected range, and previous results have been satisfactory.
3. Repeat the measurement immediately, to confirm its validity. This will determine whether the measurement was the result of a statistical fluke, a mistake or a genuine camera fault. Actions 4, 5, or 6 may then follow. This would typically apply when a result is *significantly* out of range, and previous results have been satisfactory.
4. Perform a more complex test (perhaps one usually performed monthly) to obtain more information on the nature of the problem. Actions 5 or 6 may then follow.
5. Perform a corrective action in-house (e.g. photomultiplier tube (PMT) re-tuning, acquire a new uniformity correction map) and repeat the QC test. Action 6 may also follow.
6. Call the manufacturer with a view to obtaining advice about running further tests, requesting that a service engineer run some remote diagnostics, or arranging for a planned service call-out visit if deemed necessary.

The person making the QC measurement will often be a radiographer or nuclear medicine technologist and, if all is well, they should be able to decide that the measurement is acceptable and that the camera is fit for use. However, if the measurement is not acceptable, it may require someone with more experience, such as the MPE, to deal with actions 2–6. In any case, local protocols should be in place to indicate who can take responsibility for each level of action and their decisions should be recorded in a suitable log.

Hopefully, in most cases, the appropriate decision will be that the camera is working as expected and that it can continue to be used for clinical studies. However, even this 'no action' result should be recorded by the person who performed the test signing a QC record to indicate that the camera is 'fit for clinical use'. Implicit in this decision is the fact that there is some acceptable value of each measured parameter beyond which something must be done and this is called the *action threshold.* In this section of this report, it will be assumed that we are dealing with a parameter such as uniformity or resolution where a small value is good and a large value is bad, so something only needs to be done when the parameter exceeds the action threshold. For some parameters, such as count-rate capability, where a large value is good, something would only need to be done if the parameter fell below the action threshold. In some cases, such as sensitivity, either an unexpectedly low or high value might be bad, so there would need to be both low and high action thresholds.

One way to set an action threshold would be to simply take the specified value for the parameter from the manufacturer's camera data sheet, but this has several limitations. First, some parameters (such as count-rate performance and sensitivity) are specified as 'typical' values rather than absolute standards, so not every camera will meet the specification exactly. Second, the manufacturer will quote figures calculated according to NEMA protocols and it is not always convenient to use these protocols on a daily or weekly basis. Some of the parameters recommended for regular QC in this report cannot be compared directly with the manufacturer's specification. Third, any measurement will be subject to random errors, so individual results may be above or below the 'true' value of the parameter. For these reasons, it is necessary to define an action threshold for each parameter using data acquired using the actual protocol that will be used for regular QC measurements and taking account of the expected measurement error. Thus, the action thresholds may bear no relationship with the manufacturer's data sheet for the camera. However, the purpose of acceptance testing is to ensure that the camera is performing in an acceptable manner relative to the manufacturer's specification when it is first installed. Therefore, baseline measurements using the adopted regular QC protocol should be started at the same time as acceptance testing so that initial values of these parameters can be cross-referenced with acceptance test results.

The usual way to view the results of serial QC measurements is to plot them on a graph as a time sequence. If the action threshold is plotted as a horizontal line on the same graph, this becomes known as a control graph because action needs to be taken to control the graph below the action threshold. If the daily, weekly or monthly results are entered into a spreadsheet, it becomes easy to keep a control graph of this sort up to date. The control graph can form one part of a wider process of quality control known as statistical process control (SPC). It has also been

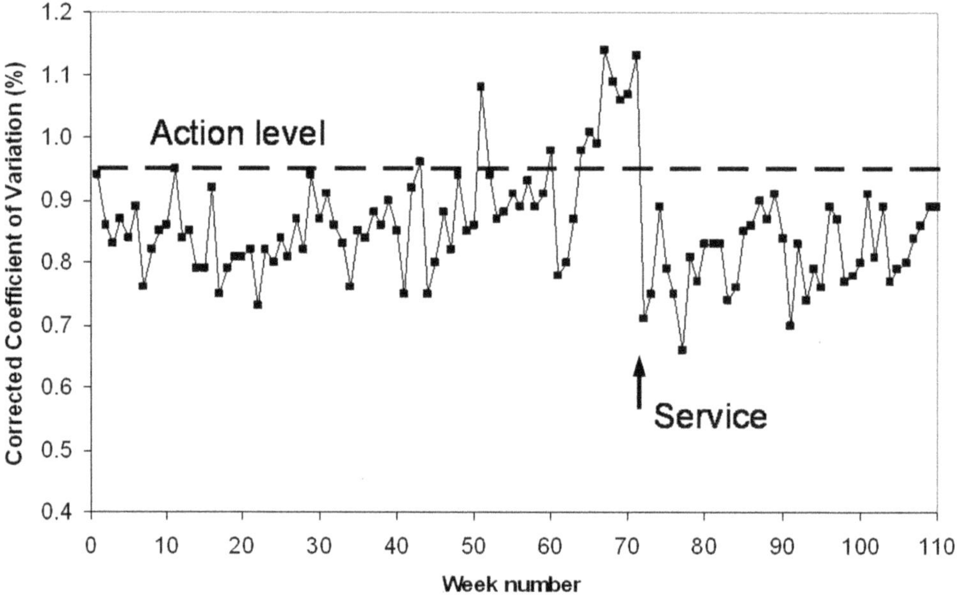

Figure 9.1 Control graph for weekly uniformity measurements of one head of a double headed gamma camera

suggested that an alternative graph showing a cumulative sum (CUSUM) analysis can be more sensitive to small changes in uniformity (Knight and Williams, 1992).

Figure 9.1 shows an example of a control graph for weekly QC of the uniformity of one head from a double headed gamma camera over a 2-year period. The data for the other head have been omitted for clarity, but in practice, results for both heads can be displayed on the same graph. The corrected coefficient of variation (Section 3.3.4) has been plotted against the measurement date. The action threshold, shown by the horizontal dashed line, was determined from the data acquired over the first 20 weeks. During this period, the camera was seen to be stable and behaving in the same way as it did at acceptance testing, so it is assumed that the fluctuations in the value of the corrected coefficient of variation simply reflect random errors in the measurement. The mean of the first 20 measurements was 0.85 with a standard deviation (SD) of 0.05. The action threshold was then set at the mean plus two standard deviations, i.e. $0.85 + 2 \times 0.05 = 0.95$. The justification for this choice of action threshold is that, if the results are normally distributed, then 95% of measurements should be within 2 SD of the mean. This implies that 2.5% of measurements are expected to fall more than 2 SD above the mean. In other words, there is a 1 in 40 chance that a value will exceed the action threshold by chance. Therefore, if a measurement exceeds the action threshold, the correct course of action is to repeat the measurement because the probability of getting two measurements above the action threshold by chance is very small. If a repeat measurement confirms that the value is still above the action threshold, then further action is required.

In Figure 9.1, it can be seen that, in week 43, the uniformity measurement was 0.96 which just exceeded the action threshold. However, after examining the acquired uniformity images, it was decided that the image was still satisfactory, so the camera was signed off as fit for clinical use with a note to check again the following week. The next week, the value was below the action threshold indicating that this had just been a random fluctuation. In week 51, the uniformity measurement gave a value of 1.08 which was well above the action threshold. The measurement was repeated immediately and found to be satisfactory, showing that there had been some error in placement of the source for the first measurement. In week 64, the measurement again exceeded the action threshold and remained high after repeating. The acquired image was examined carefully and it was judged that the area of non-uniformity was in a position that did not warrant taking the camera out of service. In fact, the uniformity got worse 2 weeks later but the camera was still judged to be usable with caution. A service visit was booked for a convenient date and the engineer fixed the problem and acquired a new sensitivity map in week 72.

In this case, something happened to the camera in week 64 to make the uniformity deteriorate, but it was decided that remedial action could wait a few weeks so as not to interfere with patient studies which had already been booked. In other circumstances, it could be that the uniformity was so bad that the camera should not be used at all for patient studies, in which case, this might entail cancelling patients if they could not be imaged on an alternative camera. However, if only one head of a double headed camera is affected in this way, then it could be possible to continue with some studies using just the good head. If the images show that the poor uniformity value is due to a problem in just one small area of the field of view, then it might still be possible to use the camera for some studies but not for others. For example, if the problem is confined to just the edge of the field of view, it might be possible to use the camera for say renography (where only the central field of view is important) but not for whole-body scans (where the whole field of view is needed). If the non-uniform area is at the side of the field of view, this might be outside the area that is used for SPECT, but if it is at the top or bottom of the field of view, it might be in an important area for SPECT, depending on the organ being imaged. For example, brain SPECT studies will usually only use one edge of a large field of view camera whereas bone SPECT may use the whole field of view. Therefore it is important to consider the appearance and location of any non-uniformity in the image as well as the quantitative value from the uniformity analysis when making decisions about whether the camera is usable or only usable for certain restricted studies. Whatever decision is made, this should be recorded and signed off by a suitably qualified member of staff.

Even when an action threshold is not exceeded, it is still important to look at the control graph to examine any trends. A slow upwards trend can give advance warning that the camera may soon need attention, so a service visit can be planned at a convenient time instead of at short notice.

It is essential that all QC measurements are recorded in a logbook or a spreadsheet and the associated images should be retained either electronically or as hard copy. Well-documented procedures, including the action to be taken when a result is obtained that is outside the control limits, are essential to the success of the QC programme. The company providing service for the camera will usually supply a record of the maintenance that they perform, but the users should also maintain a logbook for recording any problems and faults that occur during day-to-day use of the camera. This can be a valuable aid to investigating recurring problems and for assessing the camera's reliability.

9.6.3 Maintenance contracts

Most departments will wish to arrange a maintenance contract with their chosen gamma camera/nuclear medicine computer supplier and the level of ongoing support is an important factor when choosing imaging equipment. Departments tend to opt for a fully comprehensive contract if they can afford it (usually amounting to 10–15% of the purchase price per year) although different options are available:

a) Call-out at fixed hourly rate

There is no upfront financial outlay with this type of arrangement and the department will only pay for the service it actually receives. The manufacturer's service engineer will visit the site if requested but the guaranteed response time (typically 48 h) will generally not be as good as with a more comprehensive contract. The costs of spare parts will be separately charged. This may save money in the first few years after the 12-month warrantee period, but there is an obvious risk attached. Also, the possibility of switching to a more comprehensive contract when the equipment gets older and more unreliable may not be an option.

b) Preventative maintenance contract

Preventative maintenance (PM) involves the service engineer visiting the department at a set frequency (usually twice per year) to carry out a prescribed list of performance and safety checks. It will not usually include the cost of any spare parts required to perform repairs or software/safety upgrades, so this has to be budgeted separately if there is no corrective maintenance agreement.

c) Full comprehensive contracts

A fully comprehensive contract will generally include an unlimited number of priority call-outs (i.e. within 24 h), planned PM visits, and the costs of nearly all required spare parts. For nuclear medicine, the only likely (expensive) exclusion is likely to be the CT tube in SPECT/CT cameras, so a contingency budget will be required if this is the case. This type of contract is normally purchased via NHS Supplies who will have previously undertaken a national tender exercise involving the main suppliers. Other types of contract (e.g. PM only) may have to be negotiated directly with the supplier. As well as guaranteed response times, the 'premium' contract may include a specified 'up time' (the percentage of time [over a 12-month period] that the equipment was fully working and available for routine use). This figure typically would be around 95% and a discount on the following year's annual charge may be available if the equipment reliability falls below the specified level.

Another feature of some extended period (3–5 years) comprehensive contracts is the so-called 'hardware refresh' of the main computer equipment, which avoids the department having to find capital funding to replace high-resolution monitors and base units. This is important because the 'useful life' of the computer hardware will be much shorter than the gamma camera itself. This is nothing to do with the hardware becoming unreliable, but is related to the fact that computer processing power (relating to rapid advances in integrated circuit design) doubles approximately every 18 months. During the 5- to 7-year life of the gamma camera (see Section 9.7), the manufacturer will release new (generally more complex) software for SPECT reconstruction, etc. that will require increased computer power to complete routine tasks in the same time as that required by the original software. It is therefore desirable to regularly upgrade the computer hardware to keep pace with the increasing demands of complex image processing software. Typical hardware refresh will thus involve the automatic replacement of specified items (usually all workstation base units and servers, but may also include monitors) every 2–3 years, meaning that the key computers will be replaced at least twice before the camera itself is due for replacement.

d) First-line maintenance

A few Trusts choose to do their own first-line maintenance, usually after selected technical or engineering staff have attended an approved course run by the manufacturer. The idea is to filter out and fix the relatively simple faults in-house and thus only resort to calling out the manufacturer's service engineer if the fault cannot be diagnosed or fixed by local staff. These local staff may be nuclear medicine department staff, but are more usually engineering staff from another department (e.g. Clinical Engineering or Facilities). The maintenance courses (usually lasting 2–3 full days, often abroad if a multinational company) run by the manufacturer will also be expensive, so all these costs have to be factored in when comparing first-line + PM + call-out with fully comprehensive.

In all nuclear medicine departments, there will be some informal first-line maintenance by technical and scientific staff, in an attempt to keep down-time to a minimum. This does not involve the expense of formal training by local users, but can help diagnose and fix some types of problem. For example, the majority of 'communication' errors (between different elements of the system) can be cleared by a proper shut-down/restart procedure. For more complex problems, a phone conversation between the user and the service engineer will often take place before the engineer attends the department, as it may be possible to 'talk the user through' a simple corrective procedure. However, this will not normally involve removing any covers from the gamma camera itself. The engineer will also be able to login to the user's system and run some remote diagnostic software (and/or view log files), to give a clearer idea of where the problem lies. It should be noted that the manufacturer may only be willing to engage with the user in this way if a comprehensive contract is in place.

e) MES arrangements

If the Trust has leased its gamma camera equipment as part of a wider managed equipment service (MES), the MES provider will also be responsible for all aspects of equipment maintenance, which would usually include a fully comprehensive contract with the manufacturer.

In this situation, the Trust has effectively outsourced the whole business of equipment procurement and maintenance (in exchange for a large annual fee), so faults would then be reported to the MES provider, who would then liaise with the manufacturer on behalf of the user department. There are obvious reservations about these 'extended chain' arrangements but MES arrangements are becoming increasingly common within the NHS, particularly in Radiology, as capital funding to replace very expensive scanners is becoming scarcer. MES contracts can vary in scope (the Trust would specify its individual MES requirements in an OJEU advert) but the most comprehensive would cover all aspects of equipment management, even including first-line maintenance, but would exclude routine quality control (for example, see www.asteral.com).

In a MES arrangement, all equipment is upgraded as required (e.g. manufacturer recommended software/firmware upgrades) during its life, and individual scanners would be completely replaced after some specified period to ensure that the Trust's diagnostic imaging services kept pace with advancing technology. This is clearly a considerable advance over the limited 'hardware refresh' included in some conventional maintenance contracts. MES contracts vary in length, but can be up to 15 years, so represent a completely different way of managing expensive hi-tech equipment within the NHS.

Although MES contracts appear to offer financial benefits to hospital Trusts, they are relatively new to the NHS and the maintenance arrangements, in particular, have not been extensively tested against the traditional direct relationship between the user and the manufacturer.

9.6.4 Service engineer visits and QC checks

a) Before the engineer starts

Prior to an engineer undertaking work on a gamma camera, a formal handover procedure should be carried out, and this handover procedure should be detailed in the department's Local Rules. The Association of Healthcare Technology Providers for Imaging, Radiotherapy and Care (AXrEM) have produced a free "Controlled area and equipment handover document" that can be downloaded from their website (AXrEM, undated). The document refers to Controlled Areas but can be used for any room.

In addition, a partial back-up of the computer system should be carried out including acquisition (and potentially processing) protocols. This will probably be part of the engineer's job sheet procedure, but it is important to check before any work is undertaken. It is also useful to obtain a copy of the job sheet before work commences so that the impact of the work on clinical operation can be assessed.

b) When the engineer has finished

There are some basic checks that should be performed (if not already done by the engineer) before the work is accepted as satisfactorily completed.

Following a *routine* gamma camera preventative maintenance visit

Before the engineer has left the department, do the following:

- Go over the list of planned work with the engineer.
- Go through any unplanned work with the engineer.
- Check whether any work was done on datafile maintenance.
- Check whether any new image correction tables have been obtained.
- Gain sight of uniformity image(s) acquired by engineer.
- Perform basic safety checks and witness correct operation of the bed and gantry.
- Sign the engineer's visit report (a copy will normally be sent electronically whilst still on site).

Once the engineer has left the department and before clinical work is resumed, perform the following tests:

- Routine daily uniformity image.
- Specific additional checks (e.g. centre of rotation (COR)) depending on exactly what type of maintenance work was performed.

Following a *corrective* maintenance visit (work on detectors, gantry, collimator exchange, bed movement, etc.)

Before the engineer has left the department, do the following:

- Go through the list of planned work.
- Gain sight of new uniformity values (if obtained) as well as the new flood image(s).
- Confirm that the reported fault is no longer present (or check new status if fault could not be completely repaired during this visit).

Once the engineer has left the department and before clinical work is resumed, perform the following tests:

- Routine daily uniformity image.
- Specific additional checks (e.g. COR) depending on exactly what type of maintenance work was performed. If the COR checks cannot be completed before patients need to be imaged, it may be necessary to restrict work to planar imaging until the check can be performed.

Systems with an X-ray component

Additional checks are required if any work on the X-ray delivery system has been carried out, which could affect absorbed radiation dose to the patient. These include:

- The daily CT check (including a check on 'noise') *must* be performed, with consideration also given to checks on Hounsfield Units, following the guidance in Section 6.1.
- Contrast and resolution should also be assessed using a phantom (if available).
- Any new artefacts must be assessed and brought to the attention of the manufacturer or MPE for diagnostic radiology.
- Any other checks as recommended by the manufacturer or MPE for diagnostic radiology.

A complete change of tube or collimator will require more extensive tests by the MPE for diagnostic radiology. If there is any doubt with regard to CT output, then the MPE for diagnostic radiology should be consulted and radiation dose measurements carried out. This will be easier to plan, and therefore reduce down-time, if prior access to the job sheet is obtained detailing the work that would be carried out on the X-ray delivery system.

Image processing computer and network

This may relate to the gamma camera manufacturer's workstations or those of a third party connected to the gamma camera system. In this context, think about network upgrades organised by an IT department, as well as service engineer visits by the gamma camera supplier. Most nuclear medicine sub-networks depend on the main hospital network for full functionality, so it is advisable to be formally and specifically notified about these.

If a gamma camera manufacturer upgrades the operating system itself (e.g. Microsoft Windows) or its own 'system software' that sits in the layer above the basic operating system, then make sure to check (or personally verify) the following:

- That all upgraded workstations function properly (e.g. load/display patient data) after reboot.
- The network connection between the manufacturer's own workstations.
- For client applications installed on hospital PC's, check the connections/login's to the manufacturers designated server – if the software on the latter has been upgraded.
- The connection to the various workstations (DICOM nodes) from other (third party) image processing systems on the network.
- The modality work-list connection from the Radiology Information System to the acquisition workstation(s).
- Check that the Nuclear Medicine system interface to the Picture Archiving and Communications System (PACS) interface is working properly. Send a test screenshot to PACS from a native application (all clinical applications will probably use the same 'middleware' utility program for copying to PACS) and from any installed third party applications (e.g. QPS/QGS) that use a different process (e.g. spooling to an intermediate folder).
- Check the presence of online documentation for any new system software or clinical applications software and discuss staff training arrangements.

c) Sign off on the visit

Once the engineer has completed their work, and quality control checks are satisfactory, the user will be required to sign the engineer's completed worksheet. The engineer will also have completed Part 2 of the AXrEM handover document, thus officially returning the equipment to the customer. A copy of the service visit report should also be obtained. This is normally done by the engineer whilst still on site and sent as an email attachment to the designated user(s). All such reports should be filed on a shared network drive (identified in an associated written procedure), making them accessible to all authorised staff.

Once the engineer has left the department, consider some further tests, such as:

- Checking that user-written help files, etc, located on Nuclear Medicine workstations are still present.
- Checking locally created customised versions of application software ('predefines') are still present, and loadable from the host application. Test all such programs using at least one set of test data for which the results are known (i.e. that were produced by the program running under the previous environment).
- Checking locally created clinical applications (written using high level language or macros) are still working, as they may rely on core library functions that have been replaced. Test all such software using at least one set of test data for which the results are known (i.e. that were produced by the program running under the previous environment).
- Check that the back-up schedule is still working.
- Check that all printers and other peripheral devices are still working, especially for operating system updates.

Local checklists should be produced (one to be completed before the engineer leaves) that can be signed and filed after every visit. Part 3 of the AXrEM form should also be signed by the user to confirm that necessary checks (refer to checklist) have been completed and the equipment can be returned to clinical use.

Engineer visits (especially if associated with software upgrades) represent an excellent opportunity for the users to learn more about their system. Unfortunately, they also provide ample opportunity for poor communication if the engineer is simply left to his/her own devices, sometimes resulting in one problem being fixed but another being created. It is thus a good idea to designate an experienced clinical scientist to spend at least some of the time overseeing the engineer's activity. Service reports usually do not contain much detail, so there is a need to liaise directly so that any new uniformity correction maps, etc, created by the engineer are properly recorded. Even when a copy of the job sheet is provided, the engineer may do extra work depending on problems encountered whilst on site.

9.7 End of camera life

According to the American Hospital Association Health Data Management Group, the expected useful life of a gamma camera is 5 years (AHA, 2008). In the UK, the

National Health Service is more likely to use a figure of 7 years. Therefore, this report recommends that replacement of a gamma camera should be planned after a period of between 5 and 7 years. However, the actual time before a camera is replaced can vary as there are several reasons why a camera might reach the end of its useful life.

If the camera is leased through a finance company, then an initial lease will usually be for a period of between 5 and 7 years. When the lease comes to an end, an extension may be negotiated if desired, or it may be decided that this is the appropriate time to replace the camera. If there is already money in the department's budget for annual lease payments, then replacing the camera will just mean transferring these to the new lease, although a lease on new equipment will always be more expensive than extending an existing lease on old equipment. In these circumstances, the existing gamma camera will probably still be functioning well and the justification for replacement will be that new equipment will avoid the possibility of future unreliability and also keep the department up to date with available technology. This is the efficient way to replace equipment because the replacement date will be known well in advance and so there is plenty of time to plan the tendering and installation process.

If the camera was originally bought by the hospital, then replacing it will usually involve bidding for capital funds from some central source where one may be competing against other bids for equally important equipment. The justification for replacement may be that the existing camera is producing poor quality images, is becoming less reliable, or that it is now obsolete. In fact, if the camera has become very old, the manufacturer may issue an end of service (EoS) notice indicating that the availability of spare parts can no longer be guaranteed. This does not mean that they will no longer provide service, but that, in the event of a major component failure, they may not be able to repair it. Indeed, a department may request such a letter from the manufacturer to strengthen its case for replacement. The capital cost of a new camera can be offset against the increasing maintenance costs and risk of unexpected failure of an obsolete camera. Even when the camera is still functioning as well as it was when new, there may still be a case for replacing it if it no longer provides all of the facilities that are currently needed. For example, a double headed camera may be needed instead of a single headed one to cope with increased workload, or a hybrid SPECT-CT system may be required to replace a simple SPECT system to provide new imaging services. It is not unusual to have to keep bidding for funds for several years before becoming successful, so this may mean that the old camera has to be kept going for longer than the ideal 5 to 7 years. However, once the bid is successful, there is then hopefully sufficient time to plan for replacement by a given date, often the end of the financial year.

A third reason for wanting to replace a gamma camera might be that it has suffered a sudden failure that renders it unusable and it is not economical to repair. This is a difficult situation because even though funds for replacement can be made available quickly, the time taken to purchase and install a new camera will inevitably lead to a lengthy disruption to the nuclear medicine service. Avoiding this situation can be an important argument for scheduling replacement before the camera gets too old. The argument is strengthened if a logbook shows that the camera is becoming unreliable or service records show that it is requiring frequent repairs.

In all cases, the camera's service history, logbook, and QC record can be important factors in deciding when replacement is required. Camera down-time for routine service visits is necessary in order to keep the equipment working optimally, but if the service record and logbook show increasing additional down-time to rectify unexpected faults, then this can be a sign that the camera needs replacing because of unreliability. Likewise if the control graph of QC measurements shows that the camera is frequently exceeding action thresholds and requiring adjustment time to bring it back to within specification, this is also an indication of unreliability. A slowly rising control graph can also give an indication that the camera is gradually deteriorating with time. It may be that the camera is regularly exceeding the action threshold but that the manufacturer is not able to rectify this. This is clearly good evidence that the camera needs replacing, but it probably also means that the action threshold should be adjusted to reflect the current (inferior) performance of the camera because there is no point in trying to take action every week if experience shows that there is nothing that can be done about it.

Even when the control graph shows that the camera is reliable and that there is no long-term deterioration with time, it may be that the values obtained are not as good as those that could be achieved with a new camera because of improvements in technology. However, it may not be easy to obtain comparable figures for a new camera unless one can be sure that they were acquired and calculated in the same way. Moreover, the daily or weekly control graph is likely to measure just camera uniformity. Although uniformity is a good parameter to record regularly because it is sensitive to most things that might go wrong with the camera, it is still helpful to measure further parameters from time to time. This is where the annual camera QC checks can be helpful in determining whether parameters such as spatial resolution and energy resolution (Chapter 3) as well as reconstructed SPECT uniformity and SPECT performance (Chapter 5) have deteriorated since the camera was new.

Once it has been agreed that the time has come to replace the camera for any of the above reasons, the cycle begins again by commencing the procurement process for a new camera as described at the beginning of this chapter.

Appendix A References

AAPM (1995) *Quantitation of SPECT Performance*. Report No 52. American Association of Physicists in Medicine, New York. Available online at http://www.aapm.org/pubs/reports/rpt_52.pdf [accessed on 29/10/13].

AHA (2008) *Estimated Useful Lives of Depreciable Hospital Assets*. American Healthcare Association, Chicago, USA.

ARSAC (2010) The Administration of Radioactive Substances Advisory Committee. https://www.gov.uk/government/organisations/administration-of-radioactive-substances-advisory-committee [accessed on 07/01/2015].

AXrEM (undated) *Introduction and Rationale for the Development of the AXrEM Radiation Controlled Area and Equipment Handover Document.* Association of Healthcare Technology Providers for Imaging, Radiotherapy and Care, London. Available from http://www.axrem.org.uk/radiation_safety.html [accessed on 03/06/2014].

Beauregard, J.M., Hofman, M.S., Pereira, J.M., Eu, P. and Hicks, R.J. (2011) Quantitative (177)Lu SPECT (QSPECT) imaging using a commercially available SPECT/CT system. *Cancer Imaging*, Vol. 11, 56–66.

Benedetto, A. and Nusynowitz, M. (1977) An Improved FORTRAN program for calculating modulation transfer functions: concise communication. *Journal of Nuclear Medicine*, Vol. 18, 85–86.

Blokland, J.A., Camps, J.A. and Pauwels, E.K. (1997) Aspects of performance assessment of whole body imaging systems. *European Journal of Nuclear Medicine*, Vol. 24, 1273–1283.

BNMS (undated) *Standards of Delivery of a Nuclear Medicine Service.* http://www.bnms.org.uk/procedures-guidelines/bnms-generic-guidelines/standards-of-delivery-of-a-nuclear-medicine-service.html [accessed on 28/10/13].

BNMS (2013) *Organisation Audit Relaunch 2013*. British Nuclear Medicine Society. Available from: http://www.bnms.org.uk/resources/organisational-audit.html [accessed on 13/11/13].

Bolster, A.A., Waddington, W.A., *et al.*, (1996) *Gamma Camera Performance: Technical Assessment Protocol: Report*. Medical Devices Agency. Gamma Camera Assessment Team, London.

Britten, A.J., Evans, W.D., *et al.*, (2005) *Guidelines on the Quality Assurance of Intraoperative Gamma Probes*. UK Gamma Probe Working Group. Available from: http://www.bnms.org.uk/images/stories/downloads/pdf/guidelines/draft_probeqav1_sep05.pdf [accessed on 31/10/2013].

BSI (1998) BS EN 61675-3. *Radionuclide Imaging Devices. Characteristics and Test Conditions. Gamma Camera Based Wholebody Imaging Systems*. British Standards Institute, London.

BSI (2005) BS EN 60789:2005. *Medical Electrical Equipment. Characteristics and Test Conditions of Radionuclide Imaging Devices. Anger Type Gamma Cameras*. British Standards Institution, London.

Buckley, S.E., Saran, F.H., Gaze, M.N., Chittenden, S., Partridge, M., Lancaster, D., Pearson, A. and Flux, G.D. (2007) Dosimetry for fractionated (131)I-mIBG therapies in patients with primary resistant high-risk neuroblastoma: preliminary results. *Cancer Biotherapy and Radiopharmaceuticals*, Vol. 22, 105–112.

Busemann Sokole, E. (1990) Software quality control. In: *Quality Assurance in Nuclear Medicine Imaging*. Rodopi, Amsterdam, 205–228.

Busemann Sokole, E., Heckenberg, A. and Bergmann, H. (1996) Influence of high-energy photons from cobalt-57 flood sources on scintillation camera uniformity images. *European Journal of Nuclear Medicine*, Vol. 23, 437–442.

Busemann Sokole, E., Plachcinska, A., Britten, A.; EANM Physics Committee (2010a) Acceptance testing for nuclear medicine instrumentation. *European Journal of Nuclear Medicine and Molecular Imaging*, Vol. 37, 672–681.

Busemann Sokole, E., Plachcinska, A., Britten, A.; EANM Working Group on Nuclear Medicine Instrumentation Quality Control, Lyra Georgosopoulou, M., Tindale, W. and Klett, R. (2010b) Routine quality control recommendations for nuclear medicine instrumentation. *European Journal of Nuclear Medicine and Molecular Imaging*, Vol. 37, 662–671.

Chang L.T. (1978) A method for attenuation correction in radionuclide computed tomography. *IEEE Transactions on Nuclear Science*, Vol. NS-25, No. 1, 638–643.

Cox, N.J. and Diffey, B.L. (1976) A numerical index of gamma-camera uniformity. *British Journal of Radiology*, Vol. 49, 734–735.

EANM (undated) UEMS/EBNM Committees – Accreditation process. http://uems.eanm.org/index.php?id=46 [accessed on 28/10/13].

Erlandsson, K., Kacperski, K., van Gramberg, D. and Hutton, B.F. (2009) Performance evaluation of D-SPECT: a novel SPECT system for nuclear cardiology. *Physics in Medicine and Biology*, Vol. 54, 2635–2649.

GE Healthcare (2008) *Infinia + Hawkeye 4 Operator Manual, Nuclear Medicine Imaging System*. General Electric Company, Milwaukee, Wisconsin, USA.

Gibson, C.J. (1992) Gamma-camera linearity and resolution assessed using coarsely sampled line spread functions. *Physics in Medicine and Biology*, Vol. 37, 371–377.

Hart, G.C. (1986) Moire interference in gamma camera quality assurance images. *Journal of Nuclear Medicine*, Vol. 27, 820–823.

Hillel, P., Hanney, M., Redgate, S., Taylor, J. and Randall, D. (2011) Assessing the performance of a solid-state cardiac gamma camera prior to its introduction into routine clinical service. *Journal of Nuclear Medicine*, Vol. 52 (Supplement 1), 1937.

HPA (1978) *The Theory, Specification and Testing of Anger Type Gamma Cameras*, edited by P.W. Horton. Topic Group Report 27. The Hospital Physicists' Association, London.

Hughes, A. and Sharp, P.F. (1989) The sensitivity of objective indices to changes in gamma camera non-uniformity. *Physics in Medicine and Biology*, Vol. 34, 885–893.

Hutton, B.F. (2010) New SPECT technology: potential and challenges. *European Journal of Nuclear Medicine and Molecular Imaging*, Vol. 37, 1883–1886.

IAEA (2003) *IAEA Quality Control Atlas for Scintillation Camera Systems*. International Atomic Energy Agency, Vienna. Available from http://www-pub.iaea.org/MTCD/publications/PDF/Pub1141_web.pdf [accessed on 28/10/2013].

IAEA (2009) *Quality Assurance for SPECT Systems*. IAEA Human Health Series No 6. International Atomic Energy Agency, Vienna. Available online at http://www-pub.iaea.org/MTCD/publications/PDF/Pub1394_web.pdf [accessed on 29/10/13].

IEC (2005) IEC 60789. *Medical Electrical Equipment. Characteristics and Test Conditions of Radionuclide Imaging Devices. Anger Type Gamma Cameras*. International Electrotechnical Commission, Geneva.

IEC (2010) *IEC 80001-1:2010 Application of Risk Management for IT Networks Incorporating Medical Devices – Part 1: Roles, Responsibilities and Activities*. International Organisation for Standardisation, Switzerland.

IPEM (2003a) *Quality Control of Gamma Camera Systems*, edited by A. Bolster. Report 86. Institute of Physics and Engineering in Medicine, York, UK.

IPEM (2003b) *Measurement of the Performance Characteristics of Diagnostic X-ray Systems Used in Medicine. Computed Tomography X-ray Scanners. Report 32* Part III. Institute of Physics and Engineering in Medicine, York, UK.

IPEM (2004) Report 88. *Guidance and Use of Diagnostic Reference Levels for Medical X-ray Examinations*. Institute of Physics and Engineering in Medicine, York, UK.

IPEM (2005) *Recommended Standards for the Routine Performance Testing of Diagnostic X-Ray Systems*. Report 91. Institute of Physics and Engineering in Medicine, York, UK.

IPEM (2010) *Recommended Standards for the Routine Performance Testing of Diagnostic X-Ray Systems*. Report 91 (2nd edition). Institute of Physics in Engineering and Medicine, York, UK.

IPEM (2012) *The Critical Examination of X-Ray Generating Equipment in Diagnostic Radiology*. Report 107. Institute of Physics in Engineering and Medicine, York, UK.

IPEM (2013) *Quality Assurance of PET and PET/CT Systems*, edited by L. Pike. Report 108. Institute of Physics and Engineering in Medicine, York, UK.

IPSM (1992) *Quality Control of Gamma Cameras and Associated Computer Systems*, edited by J. Hannan. Report No 66. The Institute of Physical Sciences in Medicine, York, UK.

IRMER (2000) *The Ionising Radiation (Medical Exposures) Regulations 2000*. SI 2000, No 1059. HMSO, London.

IRR (1999) *The Ionising Radiation Regulations 1999*. SI 1999, No 3232. HMSO, London.

ISCORN (2011) *Database of Dynamic Renal Scintigraphy*. Available from: http://www.dynamicrenalstudy.org/ [accessed on 11/11/2013].

ISO (2008) ISO 9001. *Quality Management Systems – Requirements*. International Organization for Standardization, Switzerland.

Jarritt, P.H., Whalley, D.R., Skrypniuk, J.V., Houston, A.S., Fleming, J.S. and Cosgriff, P.S. (2002) UK audit of single photon emission computed tomography reconstruction software using software generated phantoms. *Nuclear Medicine Communications*, Vol. 23, 483–491.

Kacperski, K., Erlandsson, K., Ben-Haim, S. and Hutton, B.F. (2011) Iterative deconvolution of simultaneous ^{99m}Tc and ^{201}Tl projection data measured on a CdZnTe based cardiac SPECT scanner. *Physics in Medicine and Biology*, Vol. 56, 1397–1414.

Knight, A.C. and Williams, E.D. (1992) An evaluation of cusum analysis and control charts applied to quantitative gamma-camera uniformity parameters for automated quality control. *European Journal of Nuclear Medicine*, Vol. 19, 125–130.

Koniklijke Philips N.V. (2014) *Instructions for Use: NM Application Suite on the IntelliSpace Portal*. Philips Medical Systems Nederland B.V., Veenpluis 4-6, 5684 PC Best, The Netherlands.

Lawson, R.S. (2011) SPECT reconstruction. In: *Mathematical Techniques in Nuclear Medicine*, edited by C. Grieves. IPEM Report 100. Institute of Physics and Engineering in Medicine, York, UK.

Lawson, R.S. (2013) *The Gamma Camera. A Comprehensive Guide*. Institute of Physics and Engineering in Medicine, York, UK.

MARS (1978) *The Medicines (Administration of Radioactive Substances) Regulations 1978*. SI 1978, No 1006. HMSO, London.

MDA (2001) *Gamma Camera Performance Assessment Protocol. Part 2, Wholebody Imaging Systems*. Medical Devices Agency, Gamma Camera Assessment Team, London.

Murray, A.W., Barnfield, M.C. and Thorley, P.J. (2014) Optimal uniformity index selection and acquisition counts for daily gamma camera quality control. *Nuclear Medicine Communications*, Vol. 35(10), 1011–1017.

Neff, R., Hoops, D. and Simmons, G. (1977) A modified FORTRAN program for the calculation of modulation transfer function. *Journal of Nuclear Medicine*, Vol. 18, 1238-1240.

NEMA (2012) *Performance Measurements of Gamma Cameras*. NEMA Standards Publication NU 1-2012. National Electrical Manufacturers Association, Rosslyn, VA.

NRPB (2005) NRPB-W67 *Doses from Computed Tomography (CT) Examinations in the UK – 2003 Review*. National Radiological Protection Board, Didcot. Available from: http://www.hpa.org.uk/webc/HPAwebFile/HPAweb_C/1194947420292 [accessed on 30/10/2013].

Nuclear Fields (2013) *Collimators for Nuclear Medicine: Design Calculator*. Available from: http://www.nuclearfields.com/collimators-design-calculator.htm [accessed on 21/05/2014].

Ritt, P., Vija, H., Hornegger, J. and Kuwert, T. (2011) Absolute quantification in SPECT. *European Journal of Nuclear Medicine and Molecular Imaging*, Vol. 38, Suppl 1, S69–77.

Seo, Y., Wong, K.H., Sun, M., Franc, B.L., Hawkins, R.A. and Hasegawa, B.H. (2005) Correction of photon attenuation and collimator response for a body-contouring SPECT/CT imaging system. *Journal of Nuclear Medicine*, Vol. 46, 868–877.

Wambersie, A. (2008) *Radiation Quantities and Units, Dose to the Patients, and Image Quality in Computed Tomography (CT) (RAD UNITS)*. Final Report. European Commission EUR 23464. Available from: ftp://ftp.cordis.europa.eu/pub/fp6-euratom/docs/rad_units_projrep_en.pdf [accessed on 23/03/2015].

Wasserman, H.J. (1998) Quantification of gamma camera spatial resolution by means of bar phantom images. *Nuclear Medicine Communications*, Vol. 19, 1089–1097.

Yoneda, H., Shirao, S., Koizumi, H., Oka, F., Ishihara, H., Ichiro, K., Kitahara, T., Iida, H. and Suzuki, M. (2012) Reproducibility of cerebral blood flow assessment using a quantitative SPECT reconstruction program and split-dose 123I-iodoamphetamine in institutions with different gamma-cameras and collimators. *Journal of Cerebral Blood Flow & Metabolism*, Vol. 32, 1757–1764.

Zeintl, J., Vija, A.H., Yahil, A., Hornegger, J. and Kuwert, T. (2010) Quantitative accuracy of clinical 99mTc SPECT/CT using ordered-subset expectation maximization with 3-dimensional resolution recovery, attenuation, and scatter correction. *Journal of Nuclear Medicine*, Vol. 51, 921–928.

www.ingramcontent.com/pod-product-compliance
Ingram Content Group UK Ltd.
Pitfield, Milton Keynes, MK11 3LW, UK
UKHW061437070726
13610UKWH00019B/106